Dawson's
£11.95
23/2/15

Oxford
STUDENT'S
Science
DICTIONARY

Chris Prescott

To Ann, Peter and David

OXFORD
UNIVERSITY PRESS

D0569505

Introduction

This study dictionary has been written for students in the 11–16 age group, especially those studying Key Stage 3 and Key Stage 4 Science. It should be a useful reference book for all sciences to GCSE level, including the separate sciences of Biology, Chemistry and Physics.

The dictionary has been divided into 130 scientific themes, each contained on a double-page spread. These have been arranged alphabetically. The themes have been further divided into over 2500 words and phrases, each of which is defined within a sentence or a diagram. Remember that definitions only help you towards making use of a scientific word or phrase. No definition is ever completely correct; it is only a guide. You will need to see the links made between words and begin to use the words yourself in a range of contexts before you can really say you have understood their meanings. The definition should be read through several times and any scientific words you do not understand should be looked up in the Wordfinder at the beginning of the book.

During the writing of this dictionary I have received extensive advice from the staff of Oxford University Press and their readers. Dr Jeremy Marshall gave specialist help as an editor and lexicographer. I wish to express my sincere thanks.

Chris Prescott

April 1999

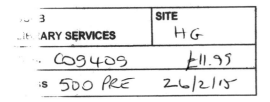

How to use this book

1. Look up the word in the Wordfinder at the front of the book. It looks like this and will give you the page number(s) you need in the Dictionary.

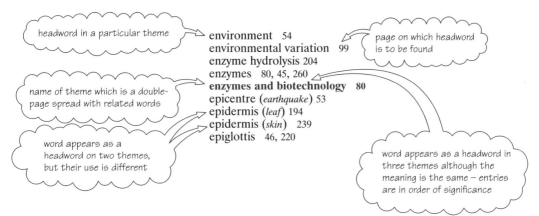

headword in a particular theme

environment 54
environmental variation 99
enzyme hydrolysis 204
enzymes 80, 45, 260
enzymes and biotechnology 80
epicentre (*earthquake*) 53
epidermis (*leaf*) 194
epidermis (*skin*) 239
epiglottis 46, 220

page on which headword is to be found

name of theme which is a double-page spread with related words

word appears as a headword on two themes, but their use is different

word appears as a headword in three themes although the meaning is the same – entries are in order of significance

2. Look up the page number(s). All the entries are in two-page spreads belonging to a theme of related words. Many of the entries will look like this:

headword being explained

alternative name for headword

first sentence is a definition of the headword

platelets (or **thrombocytes**) are very small disc-shaped cell fragments found in the blood of mammals which are important in **blood clotting**. *They are formed from larger cells with no nuclei which are made in the* **bone marrow**.

further explanation of headword is in italics

other related headwords

OR

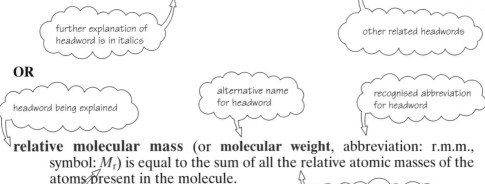

headword being explained

alternative name for headword

recognised abbreviation for headword

relative molecular mass (or **molecular weight**, abbreviation: r.m.m., symbol: M_r) is equal to the sum of all the relative atomic masses of the atoms present in the molecule.

recognised symbol for headword

first sentence is a definition of the headword

3. Look at any relevant diagrams, tables or equations nearby which will help you in your understanding of the headword. For example, there is a table on calculating relative molecular mass.

4. Some words (e.g. Silurian, Physics) appear only on the inside covers. These are not mentioned in the Wordfinder.

Wordfinder

A

abiotic factors   54
abortion   125
absolute scale (*temperature*)   113
absolute zero   113
absorption (*food*)   44
absorption of radiation   115
a.c.   68, 226
a.c. generator   66
acceleration   157, 92
acceleration of gravity   157
accommodation   87
accumulator (*secondary cell*)   12
accuracy   250
acetic acid   177
achromatic lens   136
acid hydrolysis   204
acid radical   224
acid rain   199, 260
acidic oxides   2
acids   2
acids, bases, and alkalis   2
acquired characteristics   83
actin   233
actinoids, actinide metals   183
action of a detergent   259
activation energy   22, 212
activator (*catalyst*)   213
active immunity   17
active site   80
active transport, active uptake   193
activity series   214
adaptation   82
addition polymerisation   202
addition reaction   175
additive mixing (*colour*)   137
adenine   96
adenoids   15
adenosine triphosphate   23
ADH   119, 226
adipose tissue   88, 239
adolescence   124
adrenal glands   118
adrenalin   119
aeration (*sewage*)   81
aerial (*antenna*)   36
aerobic respiration   220
aerosol   151
afferent neurones   164
afforestation   38
afterbirth   123
agglutination   17
agranulocytes   15
AIDS, acquired immune deficiency system   49
air   4
air and oxygen   4
air pollution   198
alcohol abuse   48

alcoholism   48
alcohols   176
alimentary canal   46
alkali metals   184, 185
alkaline cell   13
alkaline-earth metals   186, 187
alkalis   2
alkanes   174
alkenes   174
alkyl group   174
allele, allelomorph   96
allotropes   249
alloy   146
alloys of aluminium   146
alloys of iron   147
alluvial deposits   105
alpha decay   210
alpha particle (α)   208
alpha radiation   208
alpha sulphur   167
alternating current   68
alternator   66
altimeter   206
aluminium extraction   144
alveoli, alveolus   220
AM   36
amalgams   147
amino acids   205
ammeter   63
ammonia gas   168, 128
amniocentesis   122
amnion, amniotic sac   122
amniotic fluid   122
amoeba   29
amoebic dysentery   11
amorphous   166
amp hour   60
ampere, amp. (A)   62, 252
amphetamine   48
amphibians, Amphibia   34
amphoteric   2
amplitude (*a*)   158, 258
amplitude modulation   36
a.m.u   9
amylases   45
anabolism   116
anaerobic respiration   220
analgesic   48
analogue reading   251
analogue signals   37
androecium   197
androgens   118, 119
aneroid barometer   206
angina pectoris   17
angiosperms   31
angle of declination   143
angle of incidence (*i*)   132
angle of inclination, angle of dip   143
angle of reflection (*r*)   132

Wordfinder

Wordfinder

Wordfinder

Wordfinder

Wordfinder

Wordfinder

Wordfinder

Wordfinder

R

Wordfinder

rods (*eye*) 86
room temperature and pressure, r.t.p. 181
root hair 193
root mean square value, r.m.s. value 368
roots 192
roughage 89
round window (*ear*) 51
rounding (*of numbers*) 251
rubidium 184
runner 196
Rutherford 227
rusting 40

S

sacrificial protection 41
sacrum 234
saliva 46
salivary amylase 45
Salter's Duck 95
saltpetre 147, 184
salts 224
sand 230
sandstone 103
sandy soil 105
saponification 205
saprophytes, saprotrophs 56
satellites in orbit 159
saturated molecules 175
saturated solution 150
Saturn 241
scalar quantity 252
scapula (*shoulder blade*) 234
scientists and a few of their achievements 226
scintillation counter 210
sclerotic, sclera 86
screened methyl orange 3
screw (*Archimedes*) 226
sea salt 225
seasons 190
sebaceous glands 239
second 252
second filial generation 98
second-order lever 139
secondary atmosphere 5
secondary cell (*battery*) 12
secondary coil (*transformer*) 67
secondary colours 137
secondary consumers 56
secondary waves (*seismic*) 53
sedative 48
sedimentary rock 102
sedimentation 81
seed 217
seed dispersal 217
seed leaf 217
seismic waves 53
selective breeding 101
selective discharge 71
selective reabsorption 85

selectively permeable membrane 193
self-pollination 216
semen 120
semicircular canals 51
semiconductor 183
semilunar valves 109
semimetals 183
seminal vesicle 120
semipermeable membrane 193
sense organs, sensory organs 228, 229
sense organs in the skin 239
sensitivity 228, 18
sensory neurones 164
sepal 196
separating funnel 152
series circuit 63
serum 17
sewage 81
sex cell 219
sex chromosomes 100
sex hormones 118
sex linkage 100
sexual intercourse 124
sexual reproduction 216, 218
sexually transmitted diseases 124, 226
shadow 130
shadow zone 53
shape of the Earth 190
shooting star 241
short-sightedness 87
shoulder blade 234
shoulder girdle 235
SI units, Système International d'Unités 252, 226
sickle cell anaemia 100
sievert (Sv) 211
significant figures (sf) 251
silica 230
silicates 230
silicon, Si 230
silicon and silicates 231
silicon chip 77, 230
silicon dioxide 230
silicones 231
simple cell (*battery*) 12
simple distillation 152
sinew 233
single covalent bond 20
sinking 43
siphon 207
skeleton and muscles 232
skeleton (bones and teeth) 236
skeleton (human) 234
skin 238, 116
skin cancer 239
skull 235, 236
slaked lime 147
slate 103
slide projector 135

Wordfinder

Wordfinder

The
Dictionary

acids, bases, and alkalis

acids are **chemical compounds** which produce hydrated **hydrogen ions** H^+(aq) when in aqueous solution. *All acids, when in aqueous solution, have certain general properties. They:*

CORROSIVE

- turn moist **litmus** paper from blue to red in colour
- have a **pH** value of less than 7
- are **electrolytes** because in solution they contain ions
- react with metals like zinc and iron to form a **salt** and **hydrogen gas**
- react with metal carbonates to form a **salt**, water, and carbon dioxide gas
- neutralise **bases** and **alkalis** to form a **salt** and water.

mineral acids are **acids** which are often strong and corrosive **inorganic** compounds. *Most do not occur naturally but are made for laboratory and industrial use. They include sulphuric acid* H_2SO_4, *nitric acid* HNO_3, *hydrochloric acid HCl, and carbonic acid* H_2CO_3.

organic acids are naturally occurring **acids** found in vegetables, fruit and other foodstuffs. *They are usually weaker acids but still have a sharp or sour taste. They include ethanoic acid (found in vinegar), citric acid (in lemons), lactic acid (in milk), and oxalic acid (in rhubarb).*

acidic oxides are **oxides** of **non-metallic elements** which react with water to form acids. *Many non-metal oxides like sulphur dioxide and oxides of nitrogen are gases. If they are found in the **atmosphere**, they will dissolve in rainwater to form **acid rain**.*

Acidic oxide	Acid formed in water
carbon dioxide	carbonic acid
sulphur dioxide	sulphurous acid
phosphorus (V) oxide	phosphorous acid
nitrogen dioxide	nitric/nitrous acids

bases are **chemical compounds** that react with **acids** to form a **salt** and water. *Simple bases are oxides and hydroxides of metals. Complex bases include many organic compounds containing the amine group* (–NH_2), *such as those found in **DNA**.*

basic oxides are **oxides** of a **metallic elements** that will react with acids to form a **salt** and water only. *A few basic oxides dissolve in water to form **alkalis**.*

Basic oxide	Alkali formed in water
potassium oxide	potassium hydroxide
sodium oxide	sodium hydroxide
calcium oxide	calcium hydroxide

alkalis are water-soluble bases which produce hydrated hydroxide ions OH^-(aq) when in aqueous solution. *Potassium hydroxide solution KOH(aq), sodium hydroxide solution NaOH(aq), and calcium hydroxide solution* $Ca(OH)_2$(aq) *are alkalis, and so is ammonia solution* NH_3(aq). *All alkalis are soapy to touch and share several properties. They:*

CORROSIVE

- turn moist red **litmus** paper from red to blue
- have a **pH** value greater than 7
- are **electrolytes** because in solution they contain ions
- produce **ammonia gas** when warmed with ammonium salts
- neutralise **acids** to form a **salt** and water.

amphoteric describes a **chemical compound** that can act as an acid in one reaction but as a base in another (e.g. ZnO, Al_2O_3).

strength of acids The strength of an acid depends on its degree of ionisation in aqueous solution. *Strong acids like sulphuric acid are fully ionised in water. Weak acids like carbonic acid are only partially ionised in water: most of the ions formed recombine and remain as molecules. This is shown by the reversible sign in the chemical equation.*

$$H_2SO_4(l) \xrightarrow{\text{water}} 2H^+(aq) + SO_4^{2-}(aq)$$

$$H_2CO_3(l) \underset{\text{water}}{\rightleftharpoons} 2H^+(aq) + CO_3^{2-}(aq)$$

pH The pH scale is a logarithmic number scale (0 to 14) for showing the strength of an acid or alkali. *pH is an abbreviation for 'potential of hydrogen'. As the scale is logarithmic, a change of pH from 4 to 2 means that the substance is 100 times more acidic. Any pH value below 7 represents an acidic solution, and the lower the value the stronger the* **acid**. *Any pH value above 7 represents an alkaline solution, and the higher the value the stronger the* **alkali**.

neutral describes a solution which has a **pH** of exactly 7 and is neither acidic nor alkaline. *Neutral solutions have the same concentration of hydrogen and hydroxide ions. Pure water, salt water, and various organic liquids are neutral solutions.*

neutralisation is the chemical reaction between a **base** and an **acid** to form a **salt** and water.

acid	+	base	→	salt	+	water
HCl(aq)	+	NaOH(aq)	→	NaCl(aq)	+	H$_2$O(l)

indicator An acid–base indicator changes colour, reversibly, according to whether a solution is acidic or alkaline. *Many plant extracts, such as red cabbage juice, act as indicators.*

pH value	Colour of universal indicator	Strength
0		
1	red	strong acid
2		
3		
4	pink	weak acid
5	orange	
6	yellow	
7	green	neutral
8	turquoise	weak alkali
9	blue	
10	dark blue	
11		
12	violet	strong alkali
13		
14		

litmus is an **indicator** made from a lichen (a tiny plant) which turns red in acid and blue in alkaline solutions.

universal indicator is a mixture of several indicators and turns a range of colours corresponding to different **pH** values.

Indicator	Colour in	
	Acid	Alkali
litmus	red	blue
methyl orange	red	yellow
screened methyl orange	red	green
phenolphthalein	colourless	pink

buffer solution A buffer solution resists changes in pH when an acid or an alkali is added or when the solution is diluted. *Buffer solutions are normally weak acids and the salts of such acids (or weak alkalis and their salts). Living organisms are very sensitive to pH changes, so blood and tissue fluids contain important natural buffers.*

atmosphere The atmosphere is the **air** that surrounds the Earth and is held to it by gravity.

air is a mixture of gases, the most important of which are **nitrogen**, **oxygen**, and **carbon dioxide**. *The gases are extracted by liquefying the air. This is done by repeatedly compressing it, and then rapidly expanding it, which lowers its temperature. The components are then separated by* **fractional distillation**. *Air is the main source of* **oxygen**, **nitrogen**, *and* **noble gases**.

Component	Percentage by volume	Boiling point in °C
nitrogen	78.08	−196
oxygen	20.95	−183
argon	0.93	−186
carbon dioxide	0.03	−78
neon	0.0018	−246
helium	0.0005	−269
krypton	0.0001	−157
xenon	0.00001	−108

COMPOSITION OF AIR

oxygen is the most important gas in the air. *Oxygen has many uses and is essential for* **combustion** *and* **respiration**.
- Life support systems in hospitals use oxygen gas. Patients with breathing difficulties are also given oxygen gas.
- Oxyacetylene flames burn at very high temperatures (3300°C). Such flames are useful for welding and cutting metals.
- **Steel** manufacturing uses oxygen. The gas is 'blasted' into molten iron to remove any impurities. It does this by oxidising the main impurities, sulphur and carbon, into their oxides which escape as gases.
- Liquid oxygen (LOX) is used with rocket fuels. Outside the Earth's atmosphere, no fuel can burn unless an oxygen supply is available. Any fuel will burn more fiercely in pure oxygen than in air.
- High-altitude workers (climbers, pilots) and underwater workers (divers) need oxygen. The higher up in the atmosphere you go, the lower the air pressure becomes. The air becomes thinner and so contains less oxygen in a given volume (e.g. a lungful of air).

nitrogen is the main gas in the air, and takes part in the natural **nitrogen cycle**.
- Nitrogen's main industrial use is in the manufacture of ammonia gas by the **Haber process**. Ammonia is very important in the manufacture of **fertilisers** such as ammonium nitrate.
- Nitrogen is a very unreactive gas and is used as an inert atmosphere to prevent explosions, and in food packaging, where it prevents the oxidation of natural oils and reduces bacterial decay.
- Liquid nitrogen is used as a refrigerant. Its low temperature of −196°C is ideal for rapidly freezing foods such as vegetables and meat.

carbon dioxide is present in very small amounts in the **atmosphere** (0.03%), but it is very important because it is used for **photosynthesis** in plants. *Much carbon dioxide is dissolved in the seas and the lakes, or combined to form* **carbonates** *in the rocks.*

noble gases (or **inert gases** or **rare gases** or **group VIII** or **group 0 gases**) occupy around 1% of the **atmosphere,** of which most is argon. *All are colourless, monatomic gases which are extremely inert because their atoms have full outermost* **electron shells**.

- **argon** is used to fill ordinary and long-life light bulbs. It prevents the filament inside the bulb from burning out. Argon is also used to provide an **inert atmosphere** in arc-welding metals.
- **neon** is used in advertising signs because it glows red when electricity is discharged through it. It is also used in **Geiger–Müller tubes**.
- **helium** is very light and is used to inflate airships and weather balloons. It is also used in the helium–neon laser and to dilute the oxygen in aqualungs for divers.
- **krypton** and **xenon** are used in lamps in lighthouses, stroboscopic lamps, and photographic flash units.

troposphere The troposphere is the lowest layer of the **atmosphere**, in which most weather occurs. *Its average thickness is around 15 km (ranging from 7 km at the poles to 28 km at the equator). The temperature gradually decreases as we go higher in the troposphere.*

stratosphere The stratosphere is the second lowest layer of the **atmosphere**, up to around 50 km. *Temperature increases slightly in this layer, as it is heated from below by infrared radiation from the Earth's surface.*

ozone is formed in the upper atmosphere (**stratosphere**) when **ultraviolet radiation** from the sun breaks down **oxygen** molecules O_2 to form oxygen atoms which recombine as ozone O_3. *Ozone is also formed in the lower atmosphere by industrial activity, where it contributes to pollution.*

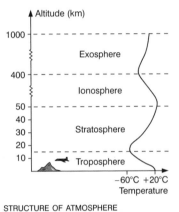

STRUCTURE OF ATMOSPHERE

ionosphere The ionosphere is the third lowest layer of the **atmosphere**, in which gases are ionised by absorption of the sun's radiation. *This layer is important for radio communication, as radio waves of certain wavelengths are reflected by its lower parts and so can travel great distances around the world.*

exosphere The exosphere is the highest region of the **atmosphere**, which begins at an altitude of about 400 km and thins out almost completely at about 1000 km.

primary atmosphere is the term for the thick cloud of gases which surrounded the Earth and other planets in our solar system when they were first formed about 4500 million years ago. *It consisted mainly of hydrogen and helium, which escaped into space.*

secondary atmosphere is the term for the mixture of gases which formed when the Earth cooled, and volcanic activity released water vapour, **methane, carbon dioxide,** and **ammonia** into the **atmosphere**. *The water vapour condensed to form the seas. There was no oxygen until the first living organisms appeared that were capable of photosynthesis, which releases oxygen. This then oxidised the ammonia and methane to form carbon dioxide, nitrogen, and more water vapour.*

atom An atom is the smallest particle of an element which can take part in a chemical reaction and remain unchanged. *These particles are extremely small. If a golf ball were magnified to the size of the Earth, then an atom would be the size of a marble! They have a radius of around 10^{-10} m and a mass of about 10^{-22} g. During chemical reactions, atoms are rearranged, but not created or destroyed (**conservation of mass**).*

nucleus (plural: **nuclei**) A nucleus is the very small central core of an atom, containing most of the atomic mass. *It is made up of **protons** and **neutrons** (except in hydrogen) and the nucleus has a positive charge. Surrounding the nucleus are 'orbiting' electrons. Any radioactive properties of an atom are associated with the nucleus. If an atom was the size of a football pitch then the nucleus (sitting on the centre spot) would be the size of a pea!*

subatomic particles are particles which are smaller than, or form part of, an **atom**. *There are three subatomic particles which make up most atoms: **protons**, **neutrons**, and **electrons**.*

Particle	Approximate radius
Atom	10^{-10} m
Nucleus	10^{-14} m
Electron	10^{-15} m

proton A proton is a positively charged **subatomic particle** which is found in the nucleus of an atom. *It has a mass equal to that of a neutron but is 1840 times heavier than an electron. Its charge is equal, but opposite, to that of an electron.*

neutron A neutron is a neutrally charged **subatomic particle** which is found in the nucleus of atoms (except hydrogen). *It has a mass roughly the same as that of a proton.*

electron An electron is a negatively charged **subatomic particle** which is found orbiting the nucleus of atoms. *Its charge is equal, but opposite, to that of a proton. However its mass is only $\frac{1}{1840}$ th that of a proton or neutron. The chemical properties of the atom are determined by these orbiting electrons, in particular the outermost ones.*

nucleon A nucleon is a general term to describe particles found in the **nucleus** of atoms. *Nucleons are either **protons** or **neutrons**.*

proton number (or **atomic number**, symbol: Z) is the number of **protons** an element has in the **nucleus** of its atom. *Every element is defined by its proton number. If you change the proton number, you change the element. The number of protons in an atom is always equal to the number of electrons, as the atom must have an overall neutral charge.*

mass number (or **nucleon number**, symbol: A) is the total number of **protons** and **neutrons** found in the nucleus of an atom. *The mass number of a particular element can vary as the number of neutrons can change.*

isotopes are atoms of the same **element** (same number of protons and electrons) with different numbers of **neutrons**, and so different mass numbers. *Nearly all elements found in nature are mixtures of several isotopes. Isotopes of a particular element have similar chemical properties but slightly different physical properties (density, rate of diffusion, etc.).*

isotopic abundance is the percentage of a particular isotope which is found in the naturally occurring element.

radioactive isotopes (or **radioisotopes**) are unstable isotopes which emit **radiation**. *Most artificial isotopes are radioactive, but most naturally occurring isotopes are not.*

electron shell An electron shell is a group of electrons that share the same 'orbit' around the nucleus of an atom. *These electrons are moving rapidly and at relatively great distances from the tiny nucleus. Each electron has its own specific orbit. The first shell can hold up to two electrons, the second shell up to eight electrons.*

energy level is the particular energy associated with an **electron shell**. *The further from the nucleus, the higher the energy of the shell.*

valence electron A valence electron is an electron found in the outermost electron shell of an atom. *The number of valence electrons determines which **group** in the **periodic table** the element belongs to. Valence electrons are important in **bonding** of atoms.*

electronic configuration (or **electronic structure**) is how the electrons are arranged in the various electron shells around the nucleus.

Element	Proton number	Mass number	Protons	Neutrons	Electrons	ELECTRONIC CONFIGURATION				Structure
						1st shell	2nd shell	3rd shell	4th shell	
hydrogen	1	1	1	0	1	1				
helium	2	4	2	2	2	2				
lithium	3	7	3	4	3	2	1			
beryllium	4	9	4	5	4	2	2			
boron	5	11	5	6	5	2	3			
carbon	6	12	6	6	6	2	4			
nitrogen	7	14	7	7	7	2	5			
oxygen	8	16	8	8	8	2	6			
fluorine	9	19	9	10	9	2	7			
neon	10	20	10	10	10	2	8			
sodium	11	23	11	12	11	2	8	1		
magnesium	12	24	12	12	12	2	8	2		
aluminium	13	27	13	14	13	2	8	3		
silicon	14	28	14	14	14	2	8	4		
phosphorus	15	31	15	16	15	2	8	5		
sulphur	16	32	16	16	16	2	8	6		
chlorine	17	35	17	18	17	2	8	7		
argon	18	40	18	22	18	2	8	8		
potassium	19	39	19	20	19	2	8	8	1	
calcium	20	40	20	20	20	2	8	8	2	

ATOMIC STRUCTURE OF FIRST TWENTY ELEMENTS

atoms and molecules

element An element is a substance that cannot be broken down into two or more simpler substances by chemical means. *Atoms of one particular element have the same number of* **protons**. *There are 92 naturally occurring elements, and each is given a* **chemical symbol**.

chemical symbol The chemical symbol represents one atom of a particular **element**. *The first letter is always a capital letter and, if there is a second, it is a small letter. Some symbols are derived from the Latin name.*

molecule A molecule is the smallest part of a substance that can take part in a chemical reaction. *A molecule is the fundamental unit of a volatile chemical compound, and consists of a group of atoms which are held together in fixed proportions by chemical bonds. All molecules have a* **chemical formula** *and many molecules contain atoms of more than one element. These are molecules of* **chemical compounds**.

chemical compound A chemical compound is a substance that consists of two or more different elements chemically bonded together in fixed proportions. *For example, water H_2O is a compound which has two hydrogen atoms and one oxygen atom chemically bonded together. Some compounds consist of* **molecules**, *while others consist of large structures held together by* **covalent bonds** *or* **ionic bonds**.

chemical formula A chemical formula is a way of showing the proportions of elements present in a **chemical compound** using symbols for the atoms present. *Subscripts are used to show the number of atoms present (see table).*

Chemical compound	Chemical formula	Atoms present		Atomicity
oxygen gas	O_2	2 oxygen atoms		2
carbon monoxide	CO	1 carbon atom 1 oxygen atom		2
hydrogen chloride	HCl	1 hydrogen atom 1 chlorine atom		2
water	H_2O	2 hydrogen atoms 1 oxygen atom		3
carbon dioxide	CO_2	1 carbon atom 2 oxygen atoms		3
ethanol (alcohol)	C_2H_5OH	2 carbon atoms 6 hydrogen atoms 1 oxygen atom		9

atomicity is the total number of atoms in a given molecule. *For example, glucose $C_6H_{12}O_6$ has an atomicity of 24.*

diatomic molecule A diatomic molecule is formed from two atoms chemically bonded together. *Examples are oxygen gas O_2, carbon monoxide CO, and hydrogen chloride HCl.*

triatomic molecule A triatomic molecule is formed from three atoms chemically bonded together. *Examples are ozone O_3, water H_2O, and carbon dioxide CO_2.*

atomic mass unit (abbreviation: **a.m.u.**) An atomic mass unit is an arbitrary unit which is equivalent to $\frac{1}{12}$th the mass of a carbon-12 atom. *It is equal to 1.66033×10^{-27}kg. This is roughly equivalent to the mass of a hydrogen atom.*

Element	Chemical symbol	Approx A_r	Element	Chemical symbol	Approx A_r
aluminium	Al	27	magnesium	Mg	24
bromine	Br	80	nitrogen	N	14
calcium	Ca	40	oxygen	O	16
carbon	C	12	phosphorus	P	31
chlorine	Cl	35.5	potassium	K	39
copper	Cu	63.5	silicon	Si	28
hydrogen	H	1	silver	Ag	108
iodine	I	127	sodium	Na	23
iron	Fe	56	sulphur	S	32
lead	Pb	207	zinc	Zn	65

relative atomic mass (or **atomic weight**, abbreviation: **r.a.m.**, symbol: A_r) is the average mass of a large number of atoms of a particular element. *This quantity takes into account the percentage abundance of various isotopes of the element that may be present. For example, chlorine gas is 75% chlorine-35 atoms and 25% chlorine-37 atoms.*

75% ^{35}Cl 25% ^{37}Cl

$$A_r = \left(\frac{75}{100} \times 35\right) + \left(\frac{25}{100} \times 37\right)$$
$$= 26.25 + 9.25$$
$$A_r = 35.5$$

RELATIVE ATOMIC MASS OF CHLORINE

relative molecular mass (or **molecular weight**, abbreviation: r.m.m., symbol: M_r) is equal to the sum of all the relative atomic masses of the atoms present in the molecule.

CALCULATIONS OF RELATIVE MOLECULAR MASS

Molecule	Calculation	M_r
H_2O water	$(2 \times 1(H)) + (1 \times 16(O))$ $= 2 + 16$	18
$CaCO_3$ calcium carbonate	$(1 \times 40(Ca)) + (1 \times 12(C)) + (3 \times 16(O))$ $= 40 + 12 + 48$	100
$C_6H_{12}O_6$ glucose	$(6 \times 12(C)) + (12 \times 1 (H)) + (6 \times 16(O))$ $= 72 + 12 + 96$	180
$(NH_4)_2SO_4$ ammonium sulphate	$(2 \times 14(N)) + (8 \times 1(H)) + (1 \times 32(S)) + (4 \times 16(O))$ $= 28 + 8 + 32 + 64$	136
$CuSO_4.5H_2O$ copper sulphate crystals	$(1 \times 63.5(Cu)) + (1 \times 32(S)) + (4 \times 16(O)) + (5 \times (2(H) + 16(O)))$ $= 63.5 + 32 + 64 + 90$	249.5

bacteria and viruses (microorganisms)

microorganisms (or **microbes**) are organisms which are too small to be seen without the aid of a microscope. *Bacteria, viruses, fungi, and protozoans are the main types of microorganisms.*

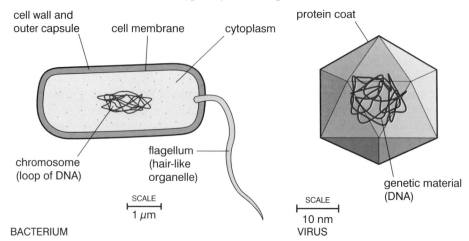

cell wall and outer capsule · cell membrane · cytoplasm · protein coat · chromosome (loop of DNA) · flagellum (hair-like organelle) · genetic material (DNA)

SCALE
1 μm
BACTERIUM

SCALE
10 nm
VIRUS

bacteria (singular: **bacterium**) are **microorganisms** which consist of a single cell without a nucleus, and with a type of cell wall unlike that of a plant cell. *Some bacteria are **pathogens** (cause disease). Others are important in ecology as **decomposers**, and in biotechnology.*

coccus A coccus is a spherical **bacterium**. *Such bacteria can join together in clumps (staphylococci) or in chains (streptococci).*

bacillus A bacillus is a rod-shaped **bacterium**. *Many are responsible for food spoilage.*

spirillum A spirillum is a rigid spiral-shaped **bacterium**.

vibrio A vibrio is a comma-shaped **bacterium**.

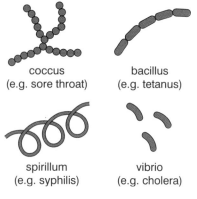

coccus (e.g. sore throat) · bacillus (e.g. tetanus)

spirillum (e.g. syphilis) · vibrio (e.g. cholera)

SHAPES OF BACTERIUM

viruses are **microorganisms** which consists of a core of nucleic acid (**DNA** or **RNA**) surrounded by a protein coat. *Viruses are about 100 times smaller than bacteria. Viruses are totally parasitic and infect plants, animals, and some bacteria (see **bacteriophage**). Outside their host organisms, viruses are inactive. Only when attached to or inside the host cell can the virus start to multiply. It interferes with the cell's normal metabolism, and the viral DNA directs the host cell to make protein coats and nucleic acid for new viruses. Many viruses are **pathogens**.*

protozoans (or **protozoa**) are single-celled **protoctists**, and unlike bacteria they have a nucleus. *Nearly all live in water, and some are **pathogens**.*

bacteria and viruses (microorganisms)

pathogen A pathogen is an organism which causes disease.

vector A vector is an agent (organism) responsible for carrying **pathogens** from one organism to another.

Pathogen	Examples of diseases
Bacteria	Diptheria, Tuberculosis, Pneumonia, Typhoid, Dysentery, Food poisoning, Cholera, Whooping cough, Yaws
Viruses	Common cold, Measles, Rubella (German Measles), Poliomyelitis, Smallpox, Influenza, Mumps, Chickenpox
Protozoans	Malaria, Amoebic dysentery
Fungi	Athlete's foot, Ringworm

malaria is a disease caused by the **protoctist** called *Plasmodium. This is carried in the saliva of the female Anopheles mosquito, which is therefore the **vector** of the disease. The mosquito sucks blood, and to stop the blood from clotting she injects some saliva. If her saliva contains Plasmodium, the person can develop malaria.*

amoebic dysentery is a disease caused by the **protoctist** called *Entamoeba. This may be present in contaminated food or drinking water. It lives in the intestine of its host and causes abdominal pain and severe diarrhoea.*

bactericide A bactericide is a substance that is used to kill harmful **bacteria**. *Common examples are **antibiotics, antiseptics,** and **disinfectants** (chemicals which kill bacteria). Sterilisation is also used to kill bacteria by heating.*

antiseptic (or **germicide**) An antiseptic is a chemical that kills or inhibits the growth of harmful microorganisms but is non-toxic to body cells.

disinfectant A disinfectant is a chemical that kills or inhibits the growth of harmful microorganisms but is toxic to body cells.

bacteriostatic describes a substance that slows down the growth and reproduction of bacteria without killing them. *Antibodies are bacteriostatic.*

bacteriophage (or **phage**) A bacteriophage is a **virus** that infects bacteria. *When inside the bacterium it can cause the bacterial cell to disintegrate. Bacteriophages are used to control manufacturing processes which use bacteria, such as **cheese making.** They are also important in **genetic engineering.***

antibiotics are substances obtained from microorganisms, especially **fungi,** that cause the destruction of other microorganisms such as disease-producing bacteria. *Common antibiotics are **penicillin,** streptomycin, and neomycin. Antibiotics do not kill **viruses.***

penicillins are a class of **antibiotics** (the commonest is penicillin G) produced from moulds of the genus *Penicillium. They kill bacteria by preventing the formation of the bacterial cell wall during the reproduction of the bacteria.*

vaccine A vaccine is a liquid preparation of treated disease-producing **microorganisms** which can stimulate the immune system to produce **antibodies** in the **blood.** *Vaccines take the form of dead or weakened bacteria or viruses that can still act as **antigens,** but cannot reproduce.*

inoculation is the introduction of a **vaccine** into the body to help fight disease.

cell A cell is a system in which two **electrodes** are in contact with an **electrolyte**. *The electrodes are normally metal or carbon (graphite). Cells produce **direct current** as a result of the **potential difference** between the two different metals and the electrolyte.*

battery A battery is a number of electric **cells** connected together. *Torch batteries often have two 1.5V cells connected in **series** to give a battery of 3V.*

voltaic cell (or **galvanic cell**) A voltaic cell is any device that produces an **electromotive force** (e.m.f.) by converting chemical to electrical energy. *The first voltaic cell was devised by Alessandro Volta (1745–1827).*

primary cell (or **simple cell**) A primary cell consists of plates of two different **metals** separated by an **electrolyte** such as salt solution or acid solution. *All primary cells have a limited life as the chemicals inside are used up and cannot be replaced. A primary cell produces electrons from the metal plate that is the more reactive (negative terminal). These electrons travel towards the metal plate that is less reactive (positive terminal). This movement of electrons produces an **electromotive force** (e.m.f.), but only for a short time because of **polarisation** and local action.*

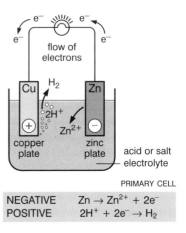

PRIMARY CELL

NEGATIVE	$Zn \rightarrow Zn^{2+} + 2e^-$
POSITIVE	$2H^+ + 2e^- \rightarrow H_2$

polarisation is the formation of bubbles of hydrogen gas on the positive plate in a **primary cell**. *These bubbles reduce the e.m.f of the cell.*

local action is the reaction of the negative plate with an acid electrolyte in a **primary cell**. *Again, hydrogen gas is produced, which reduces the electromotive force of the cell.*

dry cell A dry cell is a **voltaic cell** which does not have any free-moving liquid electrolyte. *Dry cells are used in batteries for torches, radios, etc.*

wet cell A wet cell is a **voltaic cell** which has a liquid electrolyte. *A common example is the **lead–acid accumulator**.*

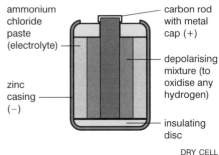

DRY CELL

secondary cell (or **accumulator** or **storage battery**) A secondary cell is a **voltaic cell** which can be recharged by connecting it to another source of electricity. *The chemical reactions taking place in a secondary cell are reversible.*

lead–acid accumulator (or **car battery**) A lead-acid accumulator is a **secondary cell** which contains electrodes made from lead and lead(II) oxide immersed in a liquid electrolyte of dilute sulphuric acid. *This is a **wet cell** which is commonly used as a car battery. It consists of six cells with an e.m.f. of 2V, so the total voltage of the battery is 12V.*

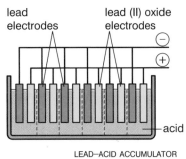

LEAD–ACID ACCUMULATOR

alkaline cell (or **nickel–cadmium cell**) An alkaline cell is a common rechargeable cell which has a negative cadmium electrode and a positive nickel electrode. *It is a **dry cell** as it has a non-liquid electrolyte of alkaline potassium hydroxide. The chemical reactions inside the cell are reversible by connecting to an outside source of electricity (recharging), so it is a **secondary cell**.*

diaphragm cell The diaphragm cell is used for the electrolysis of brine (concentrated sea water), which is a saturated solution of sodium chloride. *The diaphragm cell has two electrodes separated by a porous membrane (diaphragm). Pure brine is pumped into the cell. Electrolysis causes the negative chloride ions $Cl^-(aq)$ to be attracted to the anode, where chlorine gas is produced. This leaves a high concentration of sodium ions around the anode. Hydrogen ions $H^+(aq)$ are attracted to the cathode and form hydrogen gas. This leaves a high concentration of hydroxide ions $OH^-(aq)$ around the cathode. The sodium ions $Na^+(aq)$ are drawn through the porous membrane, where they are attracted to the $OH^-(aq)$ ions to form sodium hydroxide solution.*

DIAPHRAGM CELL

| ANODE | $2Cl^- \rightarrow Cl_2 + 2e^-$ |
| CATHODE | $2H^+ + 2e^- \rightarrow H_2$ |

fuel cell A fuel cell is a special cell in which the chemical energy of the fuel is converted directly into electrical energy. *There is no intermediate stage such as the turbine in a conventional generator. Fuel cells are more efficient than conventional generators, and produce much less pollution.*

SIMPLE FUEL CELL

NEGATIVE	$H_2 \rightarrow 2H^+ + 2e^-$
POSITIVE	$\frac{1}{2}O_2 + H_2O + 2e^- \rightarrow 2OH^-$
OVERALL	$2H^+ + 2OH^- \rightarrow 2H_2O$

13

blood is a fluid in the bodies of animals that transports oxygen and nutrients to cells, and carries waste products from the cells to the organs for excretion. *An adult human has about 5.5 litres of blood which travels around the body in the **circulatory system**. Blood has three main functions: transport, defence against disease, and the regulation of body temperature. The main components of the blood are shown in the table below.*

COMPONENTS OF THE BLOOD

Component	Size (mm)	Number (per ml)	Functions
Plasma	–	–	1 Transports carbon dioxide, food materials, hormones, and waste products (urea) in solution. 2 Liquid medium for floating blood cells, antibodies, and platelets. 3 Transports heat around the body.
White cells	0.02	7000	1 Help destroy bacteria and fight disease. 2 Make antibodies.
Red cells	0.008	5 000 000	1 Transport oxygen. 2 Transport small amount of carbon dioxide.
Platelets	0.003	250 000	1 Important in blood clotting.

plasma is the liquid part of the blood, which is about 90% water. *Floating in the plasma are the **blood cells** (red and white) and the **platelets**. Dissolved in the plasma are food materials for the body cells, waste matter (urea) and carbon dioxide, **hormones**, and **antibodies**.*

red blood cells (or **erythrocytes** or **red corpuscles**) are disc-shaped cells with no nucleus whose main function is to transport oxygen. *They are made in the **bone marrow** (each lasting approximately 120 days) and contain the red pigment **haemoglobin**. Red blood cells have flexible cell membranes so they can change shape*

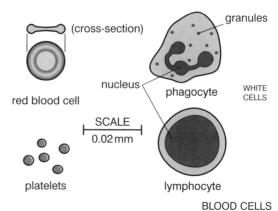

(cross-section)

red blood cell

SCALE

0.02 mm

platelets

nucleus

phagocyte

granules

WHITE CELLS

lymphocyte

BLOOD CELLS

*and flow easily through tiny blood **capillaries**. They carry oxygen to all the cells of the body, into which it passes by **diffusion**.*

haemoglobin is the blood pigment in **red blood cells** which transports oxygen. *When it combines with oxygen it becomes bright red.*

white blood cells (or **leucocytes** or **white corpuscles**) are colourless blood cells with a nucleus which are important in defence against disease. *There are two main types called **lymphocytes** and **phagocytes**.*

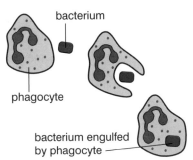

DEFENCE BY WHITE BLOOD CELLS

lymphocytes (or **agranulocytes**) are **white blood cells** made in the lymphatic tissue which produce **antibodies** to fight disease. *These are found in the lymphatic system as well as the blood. They have a large nucleus and their cytoplasm has no granules.*

phagocytes (or **granulocytes**) are large **white blood cells** which 'swallow up' foreign bodies such as bacteria (this process is called phagocytosis). *They are made in the **bone marrow** and travel by amoeboid movement. The cytoplasm of a phagocyte has tiny granules throughout.*

tissue fluid is **plasma** which has leaked from the capillaries and is returned to the blood through the **lymphatic system**.

lymphatic system The lymphatic system is a network of lymph vessels and lymph nodes (sometimes called 'glands') which helps to circulate **lymph** throughout the body.

lymph is a colourless fluid which contains the products from digestion of fats and transports both types of white cell. *The lymph eventually enters the blood system near the heart.*

lymphatic tissue (or **lymphoid tissue**) is tissue which has large numbers of lymph vessels and lymph nodes, which produce antibodies and white blood cells.

- **adenoids** are **lymphatic tissue** located at the back of the nose.
- **tonsils** are **lymphatic tissue** located in the throat.
- **spleen** The spleen is **lymphatic tissue** in the abdomen. It also stores and removes red blood cells from the blood system.

platelets (or **thrombocytes**) are very small disc-shaped cell fragments found in the blood of mammals which are important in **blood clotting**. *They are formed from larger cells with no nuclei which are made in the **bone marrow**.*

blood clotting (or **blood coagulation**) is the thickening of the blood into a clot at the site of a wound. *First the blood platelets separate and release a chemical (**thromboplastin**). This reacts with a protein in the plasma (**fibrinogen**) and causes it to harden into a fibrous substance called **fibrin**. Red blood cells and platelets get trapped in a mesh of these fibrin fibres, and so the blood clot is formed.*

haemophilia is a **genetic disease** which causes a person's blood to clot very slowly. *This is because there are too few platelets, or because the platelets are unable to release thromboplastin.*

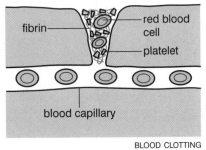

BLOOD CLOTTING

blood vessels are tubular structures through which the blood of an animal flows.

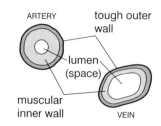

ARTERY   tough outer wall

lumen (space)

muscular inner wall

VEIN

artery An artery is a wide muscular-walled **blood vessel** that carries blood away from the heart towards the body tissue. *With the exception of the pulmonary arteries, arteries carry blood rich in oxygen (oxygenated blood) which makes them appear bright red.*

arteriole An arteriole is a small **artery** that carries blood to the capillaries.

vein A vein is a blood vessel that carries blood towards the heart and away from the body tissue. *Veins contain valves to stop the blood from flowing backwards due to gravity. This is because the blood inside is at low pressure. With the exception of the pulmonary vein, veins carry blood rich in carbon dioxide (deoxygenated blood) and waste products.*

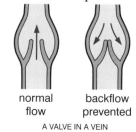

normal flow   backflow prevented

A VALVE IN A VEIN

venule A venule is a **small vein** that receives blood from the capillaries.

capillaries are the narrowest type of **blood vessel**. *Capillaries branch off from arterioles and take oxygen and dissolved food to all cells. This diffuses through the thin capillary wall into the tissue fluid surrounding each cell. Capillaries also take waste material from the cells. These capillaries join to form venules. Capillaries can be constricted or dilated according to tissue requirements.*

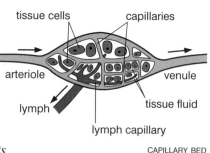

tissue cells   capillaries

arteriole   venule

lymph   tissue fluid

lymph capillary

CAPILLARY BED

blood pressure is the pressure exerted by the flow of blood through the major arteries of the body. *It is usually measured in mm Hg (millimetres of a column of mercury that the given pressure can support). Normal blood pressure is 120/80 mm Hg. The higher value is the pressure when blood is being forced out of the heart, and the lower value is the pressure when the heart muscles relax so the heart fills with blood.*

high blood pressure may be associated with various circulatory diseases:
- **thrombosis** (or **embolism**) is the formation of a small solid lump of blood (blood clot) inside a blood vessel which restricts (or stops) the flow of blood.
- **stroke** (or **apoplexy**) is caused by a cerebral thrombosis (blood clot on the brain). After a stroke, a certain area of the brain may stop working. This may result in muscles being paralysed and memory being affected.
- **arteriosclerosis** is hardening of the arteries with age as the muscular walls become less elastic, which reduces blood flow.
- **atheroma** is the build up of a fatty substance called cholesterol inside the artery walls. Such a blockage increases blood pressure.

- **heart attack** (or **coronary**) occurs when a coronary artery is blocked by **thrombosis** or **atheroma**.
- **angina pectoris** is caused by partial blockage of one or both of the coronary arteries by **atheroma** which results in severe pains in the chest.

antibodies are 'defence proteins' found in the blood plasma. *They are produced by the white blood cells called **lymphocytes** in response to **antigens** such as bacteria, toxins, viruses, etc. Some prevent the action of toxins. Others kill bacteria by dissolving their outer membranes, or cause bacteria and viruses to clump together so that they cannot reproduce properly.*

antigen An antigen is any substance from whatever source that stimulates the production of **antibodies**. *Some antigens are present in the body from birth, including those which are used in determining **blood groups**.*

blood groups provide the main method of classifying blood, according to whether the antigens A or B are present in the red blood cells. *Blood group A has antigen A only. Blood group B has antigen B only. Blood group AB has both antigens, and blood group O has neither.*

rhesus factor (or **Rh factor**) The rhesus factor is an **antigen** which may be present in blood. *If blood contains the rhesus antigen it is rhesus positive (Rh+). About 85% of the people in Britain are rhesus positive. If blood does not contain the rhesus antigen, it is rhesus negative (Rh–). Blood transfusions to Rh– people must not use blood which is Rh+. The rhesus factor was first recognised in rhesus monkeys.*

agglutination is the clumping and sticking together of red blood cells due to the reaction of antigens on their surfaces with antibodies.

blood transfusions must only be given to people of compatible blood groups whose blood can mix without **agglutination** of the red cells. *People produce antibodies against those antigens which are not normally present in their blood, and this causes agglutination.*

Blood	Can give blood to	Can receive blood from
A	A and AB	A and O
B	B and AB	B and O
AB	AB only	All groups (universal acceptor)
O	All groups (universal donor)	O only

immunity is protection of an organism against infection.

natural immunity is passed on from mother to offspring via the placenta or breast milk, or develops as a result of infection.

immunisation is the production of immunity by artificial means.

passive immunity is produced by ready-made **antibodies** injected into the body to assist in the fighting of disease (e.g. an anti-tetanus jab). *It must be 'boosted' periodically.*

active immunity is the production of **antibodies** by the body in response to the presence of antigens. *Vaccines are used to stimulate active immunity.*

serum (or **blood serum**) is blood which has had the cells and clotting substances removed but still contains specific **antibodies**.

cell A cell is a fundamental unit of living organisms. *All cells are discrete units of protoplasm surrounded by a cell membrane.*

organism (or **living organism**) A living organism is an individual living system such as an animal, plant, or microorganism. *All living organisms are made up of cells and have the following life processes: movement, reproduction, sensitivity, growth, respiration, excretion, and nutrition (remember by the mnemonic MRS GREN).*

unicellular describes living **organisms** which are made up of only one single cell, such as **bacteria** and **protozoans**.

multicellular describes living **organisms** made up of many cells. *A human being is made up of billions of cells (around 10^{14} cells).*

cell size The majority of cells in the human body are between 0.005 mm and 0.02 mm in diameter. *Such cells can be seen clearly with the use of a microscope which magnifies $400 \times$. However, not all cells are so small. An ostrich egg is a single cell and can be 20 cm long.*

protoplasm is the transparent jelly-like matter found inside living cells. *It consists of two parts: the cytoplasm and the nucleus.*

cytoplasm is the protoplasm of a living cell which is found outside the nucleus. *It is contained by a cell membrane. In the cytoplasm all the chemical reactions take place which produce energy and maintain life. These take place in tiny bodies in the cytoplasm called organelles.*

nucleus (plural: **nuclei**) The nucleus is the cell's control centre and is contained within a nuclear membrane. *It controls all the chemical reactions of the cell and also contains the genetic material of the cell.*

cell membrane The cell **membrane** forms the outer boundary of the cell. *It is through the cell membrane that exchanges between the cell and its*

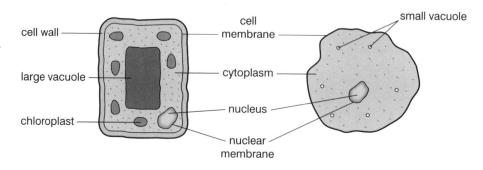

cell wall — large vacuole — chloroplast — cell membrane — cytoplasm — nucleus — nuclear membrane — small vacuole

TYPICAL PLANT CELL   TYPICAL ANIMAL CELL

Plant cell	Animal cell
Cell walls made of cellulose	Cell walls absent
Nucleus at the edge of the cell	Nucleus anywhere but often at the centre
Chloroplasts present in many cells	Chloroplasts never present
Thin lining of cytoplasm against wall	Cytoplasm throughout the cell
Often one large central vacuole	Small vacuoles throughout the cytoplasm

surroundings take place. The cell membrane is semi-permeable, which means that it is selective about which substances it allows through.

cell wall The cell wall is a rigid outer wall of plant cells, made of **cellulose**. *Because plant cells are rigid they can build on top of one another. This support allows some plants, like trees, to grow very tall.*

chloroplast A chloroplast is a tiny structure in the cytoplasm of plant cells which contains a green pigment called chlorophyll. *Chloroplasts absorb energy from sunlight and so are important for **photosynthesis**.*

vacuole A vacuole is a fluid-filled sac found in the cytoplasm of cells. *Vacuoles are small and temporary in animal cells. Plant cells have one large permanent vacuole which is filled with **cell sap**.*

cell sap is the solution of dissolved minerals and sugars that fills the large vacuoles of plants.

organelles are tiny bodies found in the **cytoplasm**.

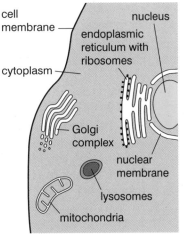

ORGANELLES (IN ANIMAL CELL)

- The **endoplasmic reticulum** is a network of membranes joined to the nuclear membrane. It has a large surface area and is involved in chemical reactions and fluid storage.
- The **ribosomes** are tiny particles attached to the endoplasmic reticulum. They are involved in the synthesis of proteins.
- The **lysosomes** are round sacs which contain powerful enzymes.
- The **mitochondria** are the organelles in the cytoplasm where energy is produced from chemical reactions.
- The **Golgi complex** collects and distributes the substances which are made in the cell.

tissue is a collection of cells which perform a specific function. *Examples are **muscle** tissue, **nerve** tissue, **skin** tissue, and **leaf** tissue.*

organ An organ is a collection of different tissues which work together to perform some function in the organism. *Examples of organs are the **skin**, **heart**, **lungs**, **brain**, and **kidneys**.*

muscle cell   gland cell

brain cell   sperm cell

nerve cell

DIFFERENT CELL SHAPES

organ system An organ system is formed by a number of different organs which together carry out a particular bodily function. *For example, the heart and blood vessels are organs which together make up the **circulatory system**. Other systems include the **urinary system**, **digestive system**, **reproductive system**, and **nervous system**.*

cell division There are three types of cell division: **binary fission** (in simple single-celled organisms), **mitosis** (in most types of cell) and **meiosis** (in sex cells only).

chemical bond A chemical **bond** is a strong force of attraction between atoms inside a molecule or crystal. *It may be an **ionic bond** or a covalent bond.*

intramolecular forces are forces between atoms inside molecules. *They typically measure around $1000\,kJ\,mol^{-1}$.*

intermolecular forces are forces between molecules. *They are much weaker than intramolecular forces.*

van der Waals' forces are **intermolecular forces** caused by induced dipoles (separation of positive and negative charge) brought about by the movement of electrons. *They typically measure around $10\,kJ\,mol^{-1}$.*

covalent bond A covalent bond is a **chemical bond** formed by the sharing between atoms of their outermost electrons. *All **molecules** contain covalent bonds. Electrons are shared so that the outermost shell of each atom has a stable inert gas configuration.*

single covalent bond A single covalent bond is the sharing between two atoms of two electrons (one from each atom).

double covalent bond A double covalent bond is the sharing between two atoms of four electrons (two from each atom).

triple covalent bond A triple covalent bond is the sharing between two atoms of six electrons (three from each atom).

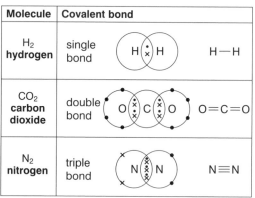

Molecule	Covalent bond		
H_2 hydrogen	single bond	H $\overset{\cdot}{\underset{\times}{}}$ H	H—H
CO_2 carbon dioxide	double bond	O C O	O=C=O
N_2 nitrogen	triple bond	N N	N≡N

COVALENT BONDING

metallic bond A metallic bond is a **chemical bond** of the type which holds together the atoms in a metal or alloy. *The atoms are packed closely together in fixed positions in a lattice. This allows the outermost electrons in the atoms to become 'delocalised'. These electrons are able to move freely through the lattice. The result is positive metal ions in a 'sea of electrons'. The existence of free electrons accounts for the high electrical and thermal conductivity of metals.*

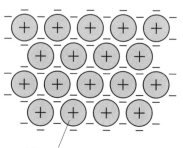

positive metal ions in a 'sea' of delocalised electrons

METALLIC BONDING

ion An ion is a charged particle formed when an **atom** (or group of atoms) gains or loses one or more **electrons**.

ionic bond An ionic bond is a **chemical bond** formed by transfer of one or more electrons from the outer shell of a metal atom to the outer shell of a non-metal atom. *The metal atom as it loses electrons becomes a positive ion. The non-metal atom as it gains electrons becomes a negative ion. The bonding is the electrostatic attraction between these oppositely charged ions. Enough electrons are transferred or gained to ensure that the outermost shell of the ion formed has a stable inert gas configuration.*

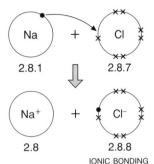

IONIC BONDING

cation A cation is a positively charged **ion** that is attracted to the **cathode** during electrolysis. *All metal ions and the hydrogen ion are cations.*

anion An anion is a negatively charged **ion** that is attracted to the **anode** during electrolysis. *All non-metal ions and most radicals (except the ammonium ion) are anions.*

Cation	Anion	Ionic compound
copper(II) Cu^{2+}	chloride Cl^-	copper(II) chloride $CuCl_2$
sodium Na^+	oxide O^{2-}	sodium oxide Na_2O
iron(III) Fe^{3+}	hydroxide OH^-	iron(III) hydroxide $Fe(OH)_3$
ammonium NH_4^+	sulphate SO_4^{2-}	ammonium sulphate $(NH_4)_2SO_4$

ionic lattice An ionic lattice is a giant structure of tightly packed ions. *It is a regular arrangement and each ion is surrounded by oppositely charged ions. Many crystalline substances have ionic lattices.*

lattice energy Lattice energy is the energy required for the complete separation of ions in one mole of an ionic substance. *It is an indication of the strength of the ionic bond.*

coordination number is the number of nearest neighbours of an atom or ion in a structure. *For the sodium chloride lattice, the coordination number of each ion is 6.*

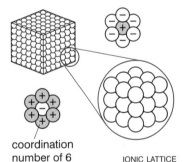

coordination number of 6

IONIC LATTICE

Ionic compounds	*Covalent compounds
1 are made of charged particles (ions)	1 are made of molecules
2 have strong chemical bonds between ions	2 have strong chemical bonds inside the molecule
3 form giant ionic lattices	3 have weak attractive forces between molecules
4 are crystalline solids	4 are often gases or volatile liquids
5 have high melting and boiling points	5 have low melting and boiling points
6 are usually water soluble	6 are often insoluble in water
7 conduct electricity when molten or aqueous	7 do not conduct electricity

(*Macromolecules are exceptions. For more information see **structures and properties**.)

chemical energy is the energy which is released or absorbed when chemical bonds are rearranged. *If stronger bonds are formed, then energy is released. Some chemical reactions produce electrical energy (see **cells**) or light energy, but nearly all reactions involve heat changes. In many chemical reactions heat is produced, and this causes a temperature rise in the surroundings. Sometimes, the opposite happens and heat energy is taken in from the surroundings. This causes a temperature drop.*

exothermic reaction An exothermic reaction is a **chemical reaction** during which heat energy is transferred to the surroundings. *All exothermic reactions are accompanied by a temperature rise in the surroundings.*

endothermic reaction An endothermic reaction is a **chemical reaction** during which heat energy is taken in from the surroundings. *All endothermic reactions are accompanied by a temperature fall in the surroundings.*

energy level diagram (or **profile diagram**) An energy level diagram is one which shows the energy change during a chemical reaction. *It is a plot of energy change against time.*

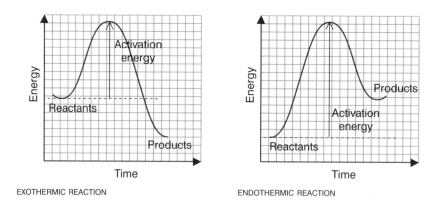

EXOTHERMIC REACTION   ENDOTHERMIC REACTION

bond making During chemical reactions, when bonds are made, energy is given out in an **exothermic** process. *If a chemical reaction is exothermic overall, then the energy released in making bonds is greater than the energy used in breaking all the bonds.*

bond breaking During chemical reactions, when bonds are broken, energy is absorbed in an **endothermic** process. *If a chemical reaction is endothermic overall, then the energy required to break the old bonds is greater than the energy released in making bonds.*

Average bond energies (kJ mol^{-1})	
H–H	436
H–O	464
H–Cl	431
C–C	348
C=C	611
C≡C	835
O=O	496
Cl–Cl	242
C–H	435
C–Cl	339

activation energy (symbol: E_A) is the minimum amount of energy required to start a chemical reaction through initial bond breaking, allowing new bonds to be made.

bond energy is the amount of energy required to break a particular bond in a compound. *Exact values for particular bonds vary slightly, as they depend upon what other atoms are in the compound. However, 'average bond energies' are commonly used for calculating energy changes.*

Hess's law states that in a particular chemical reaction the overall energy change is the same no matter which route is taken in going from reactants to products. *It was named after the Russian chemist Germain Henri Hess (1802–50).*

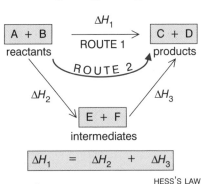

calculation of energy change is done by totalling the energy given out, by all the bonds formed, and then subtracting all the energy needed for breaking the various bonds in the reaction.

$$\Delta H_1 = \Delta H_2 + \Delta H_3$$

HESS'S LAW

energy change = sum of bonds made – sum of bonds broken

For example, consider the burning of hydrogen gas in air:

$$2H_2(g) + O_2(g) \rightarrow 2H_2O(g)$$

H–H H–H   O=O   H–O–H H–O–H

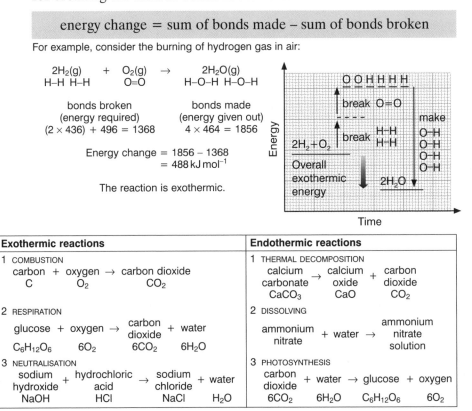

bonds broken	bonds made
(energy required)	(energy given out)
$(2 \times 436) + 496 = 1368$	$4 \times 464 = 1856$

Energy change = $1856 - 1368$
= $488 \, kJ \, mol^{-1}$

The reaction is exothermic.

Exothermic reactions	Endothermic reactions
1 COMBUSTION carbon + oxygen → carbon dioxide C O_2 CO_2	1 THERMAL DECOMPOSITION calcium → calcium + carbon carbonate oxide dioxide $CaCO_3$ CaO CO_2
2 RESPIRATION glucose + oxygen → carbon dioxide + water $C_6H_{12}O_6$ $6O_2$ $6CO_2$ $6H_2O$	2 DISSOLVING ammonium nitrate + water → ammonium nitrate solution
3 NEUTRALISATION sodium hydroxide + hydrochloric acid → sodium chloride + water NaOH HCl NaCl H_2O	3 PHOTOSYNTHESIS carbon dioxide + water → glucose + oxygen $6CO_2$ $6H_2O$ $C_6H_{12}O_6$ $6O_2$

adenosine triphosphate (or **ATP**) is a substance used as a store of chemical energy by living cells. *Removal of one of its phosphate groups releases energy for biochemical reactions.*

chemical equations

chemical reaction A chemical reaction is a change by which chemical elements or compounds rearrange their atoms to produce new chemical elements or compounds. *The total number of atoms in the reaction remains the same, so there is no change in the amount of matter. All chemical reactions involve energy changes (see **chemical energy**).*

conservation of mass is the principle that during chemical reactions atoms are rearranged, but not created or destroyed.

TYPES OF CHEMICAL REACTION

Type of reaction	Meaning	Example
Synthesis	Involves building up complicated molecules from simple ones.	**Photosynthesis** involves synthesising sugar molecules from simple molecules like carbon dioxide and water.
Decomposition	Involves breaking down compounds into simpler molecules or sometimes elements.	**Thermal decomposition** of limestone to form lime and carbon dioxide.
Combustion	Involves the chemical reaction of a substance with oxygen.	**Burning** of fuels like methane gas produces carbon dioxide and steam.
Displacement	Involves a more reactive element displacing a less reactive element.	**Reactive metals** will always displace a less reactive metal from its oxide or a solution of its salt.
Redox	Involves elements in the reaction gaining and losing electrons.	**Rusting** is a redox reaction, as iron loses electrons and oxygen gains them to form iron(III) oxide, which is rust.
Neutralisation	Involves reactions between acids and bases.	**Acids** will neutralise **bases** to form a salt and water.
Polymerisation	Involves the joining together of large numbers of molecules to form a giant molecule.	Formation of **plastics** and **artificial fibres**.
Exothermic	A reaction which involves giving out heat energy because chemical bonds have been made.	**Respiration** is an exothermic reaction as energy is released from foodstuffs.
Endothermic	A reaction which involves taken in heat energy because chemical bonds have been made.	**Dissolving** is an endothermic reaction as bonds are broken to spread the particles throughout the solution.

reactants are the chemical elements or compounds that a chemical reaction starts with. *The reactants are placed on the left-hand side in a **chemical equation**.*

products are the chemical elements or compounds that are produced during the chemical reaction. *The products are placed on the right-hand side in a **chemical equation**. The products have different properties from the reactants. Sometimes the products react with each other to re-form the original reactants. These are called **reversible reactions**.*

chemical equation A chemical equation is a way of summarising a chemical reaction. *Although it can be written in words, an equation is often written using chemical symbols and chemical formula. When you write chemical equations, to start with, it is advisable to follow these steps:*

1. Write down the equations in words, using either the information given or your own chemical knowledge.
2. Then write down the correct **chemical formula** of every reactant on the left-hand side, and every product on the right-hand side.
3. Balance the equation. This involves making sure that the number of atoms of each element is the same on both sides of the equation, so that all atoms are accounted for and none are lost or gained. Do this by changing the proportions of reactants and products, making sure that you do not change any chemical formula.
4. Finally put **state symbols** in the equation for every reactant and product. Solid is (s), liquid is (l), gas is (g), and aqueous is (aq). Aqueous means dissolved in water. Here is an example:

1.	sulphuric acid	+	sodium hydroxide	→	sodium sulphate	+	water
2.	H_2SO_4		NaOH		Na_2SO_4		H_2O
3.	H_2SO_4	+	2NaOH	→	Na_2SO_4	+	$2H_2O$
4.	$H_2SO_4(aq)$	+	$2NaOH(aq)$	→	$Na_2SO_4(aq)$	+	$2H_2O(l)$

ionic equations are used when a chemical reaction involves the coming together of ions in solution. *Consider the neutralisation of sulphuric acid with sodium hydroxide. If all the ions are written out fully we have the following:*

$$2H^+(aq) + SO_4^{2-}(aq) + 2Na^+(aq) + 2OH^-(aq) → 2Na^+(aq) + SO_4^{2-}(aq) + 2H_2O(l)$$

If we cancel out those ions which are common to both sides of the equation we get:

$$2H^+(aq) + 2OH^- → 2H_2O(l)$$

which can be simplified to the ionic equation for the neutralisation of an acid:

$$H^+(aq) + OH^-(aq) → H_2O(l)$$

calculations from chemical equations Chemical equations can be used to calculate the amounts of reactants being used and products being formed. *Consider the burning of magnesium. From the equation and the relative atomic masses of the atoms involved (Mg = 24, O = 16) we find that 48g of magnesium requires 32g of oxygen and forms 80g of magnesium oxide. The sum of the masses of the reactants must equal the sum of the masses of the product(s). Reaction between these atoms in this chemical equation is always in the same proportion by mass, which is 3 : 2 (for reactants) : 5 (for products). More is said about calculations from equations in the section on* **moles.**

$2 Mg(s)$	+	$O_2(g)$	→	$2MgO(s)$
2×24	+	2×16		$2 \times (24 + 16)$
48		32		80
48	:	32	→	80
3	:	2	→	5

reactants mass ratio | products mass ratio

eg 24g	+	16g	→	40g
96g	+	64g	→	160g
3g	+	2g	→	5g

chemical formulae

valency (or **valence**) is the combining power of an **atom** or **radical**. *In ionic compounds it is equivalent to the charge on the ion. In covalent compounds it is equal to the number of bonds formed.*

Element	Ion	Valency	Element	Ion	Valency	Radical	Ion	Valency
sodium	Na^+	1	hydride	H^-	1	ammonium	NH_4^+	1
potassium	K^+	1	chloride	Cl^-	1	hydroxide	OH^-	1
silver	Ag^+	1	bromide	Br^-	1	hydrogen carbonate	HCO_3^-	1
hydrogen	H^+	1	iodide	I^-	1			
lead(II)	Pb^{2+}	2	oxide	O^{2-}	2	hydrogen sulphate	HSO_4^-	1
copper(II)	Cu^{2+}	2	sulphide	S^{2-}	2			
magnesium	Mg^{2+}	2	nitride	N^{3-}	3	nitrate	NO_3^-	1
calcium	Ca^{2+}	2	phosphide	P^{3-}	3	sulphate	SO_4^{2-}	1
zinc	Zn^{2+}	2				sulphite	SO_3^{2-}	2
barium	Ba^{2+}	2				carbonate	CO_3^{2-}	2
iron(II)	Fe^{2+}	2				phosphate	PO_4^{3-}	3
iron(III)	Fe^{3+}	3						
aluminium	Al^{3+}	3						

VALENCY OF IONIC COMPOUNDS

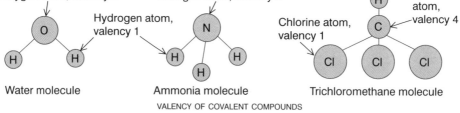

Oxygen atom, valency 2   Nitrogen atom, valency 3

Hydrogen atom, valency 1

Chlorine atom, valency 1

Carbon atom, valency 4

Water molecule   Ammonia molecule   Trichloromethane molecule

VALENCY OF COVALENT COMPOUNDS

radical A radical is a group of atoms within a compound that maintains its identity throughout a **chemical reaction**. *Such a group of atoms (unlike a molecule) cannot exist by itself.*

variable valency is the combining power of an **element** which can combine with other elements in different proportions. *Many transition metal ions can have a variable valency. To distinguish the ions we use roman numerals. For example, compounds containing Cu^+ are called copper (I) compounds to distinguish them from copper (II) compounds containing Cu^{2+}. Compounds containing Fe^{2+} and Fe^{3+} ions are distinguished as iron (II) and iron (III) compounds.*

Roman numeral	Old name
Cu(I)	cuprous
Cu(II)	cupric
Fe(II)	ferrous
Fe(III)	ferric

chemical formula A chemical **formula** is a way of showing the proportions of elements present in a **chemical compound** using symbols for the atoms present. *Subscripts are used to show the number of atoms present. This can be worked out from the valency or combining power of each atom.*

Compound	Valencies → Formula	
copper(I) oxide	Cu^1O^2	Cu_2O
copper(II) oxide	Cu^2O^2	CuO
sodium nitrate	$Na^1NO_3{}^1$	$NaNO_3$
sodium sulphate	$Na^1SO_4{}^2$	Na_2SO_4
magnesium nitrate	$Mg^2NO_3{}^1$	$Mg(NO_3)_2$
ammonium nitrate	$NH_4{}^1NO_3{}^1$	NH_4NO_3
ammonium sulphate	$NH_4{}^1SO_4{}^2$	$(NH_4)_2SO_4$

valency   valency

$$X \overset{x}{} \overset{}{Y} \overset{y}{}$$

Chemical formula X_yY_x

If $x = y$ then XY

empirical formula An empirical formula is a chemical formula that shows the simplest ratio between the atoms in a molecule.

molecular formula A molecular formula is a formula which simply gives the type and number of atoms present.

structural formula A structural formula is a formula which shows the order in which the atoms are arranged.

displayed formula A displayed formula is a formula which shows the **covalent bonding** between the atoms present.

Name	Empirical formula	Molecular formula	Structural formula	Displayed formula
ethane	CH_3	C_2H_6	CH_3CH_3	
ethene	CH_2	C_2H_4	$CH_2{=}CH_2$	
ethanol	C_2H_6O	C_2H_6O	CH_3CH_2OH	
ethanoic acid	CH_2O	$C_2H_4O_2$	CH_3COOH	

percentage composition is the make-up of a **chemical compound** expressed in terms of each of the elements present, calculated as a percentage by mass. *The table shows how to calculate the percentage composition of the elements carbon and hydrogen in various hydrocarbons.*

Name of hydrocarbon	Molecular formula	Relative molecular mass $A_r(C) = 12\ A_r(H) = 1$	Percentage composition	
			% carbon	% hydrogen
ethane	C_2H_6	$(2 \times 12) + (6 \times 1) = 30$	$\dfrac{24}{30} \times 100 = 80\%$	$\dfrac{6}{30} \times 100 = 20\%$
ethene	C_2H_4	$(2 \times 12) + (4 \times 1) = 28$	$\dfrac{24}{28} \times 100 = 85.7\%$	$\dfrac{4}{28} \times 100 = 14.3\%$
benzene	C_6H_6	$(6 \times 12) + (6 \times 1) = 78$	$\dfrac{72}{78} \times 100 = 92.3\%$	$\dfrac{6}{78} \times 100 = 7.7\%$

27

taxonomy is the study of the theory, practice, and rules of classification of living and extinct organisms. *These organisms are put in groups based on similarities in structure, appearance, **genotype**, and other characteristics.*

biological classification first classifies living organisms into very large groups called **kingdoms**. *The largest kind of subgroup within a kingdom is called a **phylum**. Each phylum is divided into **classes**, which are then subdivided into **orders**, which are divided into **families**. Every living organism belongs to a **species**. Each species belongs to a **genus**, and each genus to a family.*

phylum (plural: **phyla**) A phylum is a large group of organisms sharing a similar basic structure. A phylum of plants is often called a division.

binomial classification is the system of naming living organisms (and extinct organisms) using a two-part scientific (Latin) name. *It was devised by the Swedish botanist Carl Linnaeus (1707–78). The first part of the name is the **genus**. The second part is the **species**. For example, the binomial name for a human is **Homo sapiens**.*

Kingdom	–	Animal
Phylum	–	Chordata
Class	–	Mammalia
Order	–	Primates
Family	–	Hominidae
Genus	–	*Homo*
Species	–	*sapiens*

CLASSIFICATION OF A HUMAN BEING

genus (plural: **genera**) A genus is a group of closely related species. *Some genera contain only one species. All genera are grouped in families.*

species A species is a group containing living organisms of the same kind. *Members of a species may breed with one another, but normally cannot breed with members of another species. Rarely, very closely related species interbreed to produce a **hybrid**.*

hybrid A hybrid is the offspring of plants or animals produced from the cross of two closely related species. *For example, the hybrid of a female horse (mare) and a male donkey is a mule. Hybrids between different animals are usually sterile.*

kingdom A kingdom is the highest rank in the classification of living organisms. *Traditionally only two kingdoms, **plants** and **animals**, were recognised. In modern classification, five kingdoms are recognised including **Monera** (bacteria), **Protoctista**, and **Fungi**, as well as plants and animals.*

protozoa (or **protozoans**) are single-celled organisms with a cell membrane and a nucleus. *Some are plant-like and feed by photosynthesis (e.g. **euglena**). Others are animal-like and feed on other living things (e.g. **amoeba**, **paramecium**).*

Monera is a kingdom which includes all **bacteria**, which are unicellular (single-celled) organisms having no nucleus or normal cell wall. *The nuclear material consists of strands of DNA but it is not enclosed in a distinct nuclear membrane. The oldest fossils belong to this kingdom, so it is probable that bacteria were the first living organisms on Earth.*

prokaryotes are organisms whose genetic material is not surrounded by a nuclear membrane, such as **bacteria**.

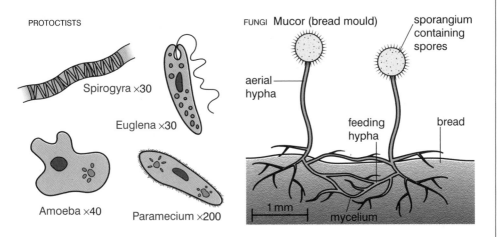

PROTOCTISTS

Spirogyra ×30

Euglena ×30

Amoeba ×40

Paramecium ×200

FUNGI Mucor (bread mould)

sporangium containing spores

aerial hypha

feeding hypha

bread

1 mm

mycelium

protoctists (kingdom **Protoctista** or **Protista**) form a kingdom which includes all unicellular (single-celled) organisms or simple multicellular (many-celled) organisms which possess nuclei but which cannot be classified as either fungi, plants, or animals. *This kingdom includes all protozoa and filamented algae such as spirogyra. Nearly all protoctists live in water.*

fungi (singular: **fungus**) are plant-like organisms which do not contain chlorophyll, and which are therefore incapable of photosynthesis. *Instead they are saprophytes or parasites and feed on organic material such as bread, faeces, dead plants or animals. Fungi mainly exist in damp conditions and are normally multicellular. Mushrooms, toadstools, and yeasts are fungi.*

mycelium is the mass of filaments which form the main part of a **fungus**.

hypha (plural: **hyphae**) A hypha is a microscopic hollow filament in a **fungus**. *Together the hyphae make up a network called the mycelium.*

sporangium A sporangium is the reproductive structure in a **fungus** which produces **spores**.

spore A spore is a reproductive cell produced by **fungi**, some plants, and some bacteria.

dichotomous keys are the simplest type of key, made up of brief descriptions arranged in numbered pairs. *You must begin at the first description and work your way through the key following the instructions. The key either names the organism, or gives you the next pair of descriptions to consult.*

1	Hair present	MAMMALS
	Hair absent	Go to 2
2	Feathers present	BIRDS
	Feathers absent	Go to 3
3	Breathe using lungs	Go to 4
	Breathe using gills	Go to 5
4	Dry scaly skin	REPTILES
	Moist scaleless skin	AMPHIBIANS
5	Fins with bony skeleton	BONY FISHES
	Fins with cartilage skeleton	CARTILAGE FISHES

plant kingdom (or **Plantae**) The plant kingdom includes all multicellular organisms which are capable of **photosynthesis**. *All plant cells have a distinct nucleus (plants are eukaryotes) and a cell wall made of cellulose, and contain* **chlorophyll** *to absorb the light energy needed for photosynthesis. The plant kingdom is divided into four main groups:* **bryophytes, ferns, conifers,** *and* **angiosperms.**

autotrophic nutrition is a process by which organisms make their own food from inorganic material. *All green* **plants** *feed this way by* **photosynthesis**, *making sugar and starch from carbon dioxide and water using the energy of sunlight.*

bryophytes are primitive plants, liverworts and mosses, with simple stems, leaves, and roots. *They live in damp conditions as they need water for reproduction. The male bryophyte produces sex cells which must swim to the female to fertilise it. After fertilisation, the female plant grows a capsule that contains* **spores**, *which when released grow to become the new plant.*

ferns are perennial flowerless plants with large leaves called fronds. *These fronds grow either from a short stem or a rhizome (underground stem) and gradually uncurl as they become more mature. Like bryophytes, ferns need water for fertilisation, but they can live in slightly drier places. They reproduce by means of spores released from capsules on the underside of their fronds.*

conifers are cone-bearing plants which produce **seeds**, but without flowers or fruits. *Most conifers are evergreen shrubs or trees like pines or firs. They can live in very dry places and their leaves are reduced to 'needles' or scales to prevent*

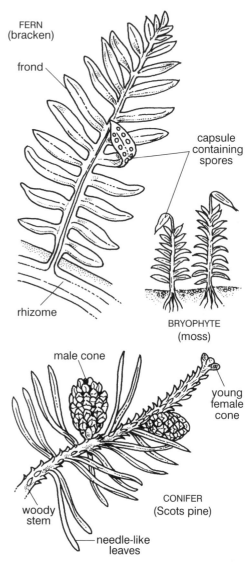

FERN
(bracken)

frond

capsule
containing
spores

rhizome

BRYOPHYTE
(moss)

male cone

young
female
cone

woody
stem

CONIFER
(Scots pine)

needle-like
leaves

excessive water loss. Unlike bryophytes or ferns, they do not need water for fertilisation. Conifers have male and female cones (their sex organs) and are usually wind-pollinated. The seeds are not protected by a fruit or carpel. The wood of conifers is softwood, which is widely used as timber in the building industry.

angiosperms (flowering plants) are seed-bearing plants that produce **flowers**. *The seeds grow inside a fruit which develops from an **ovary** inside the flower. Angiosperms are the most common plants in the plant kingdom. They are also the most highly developed and inhabit a wide range of habitats. There are two classes: **monocotyledons** and **dicotyledons**.*

monocotyledons (abbreviation: **monocots**) are a class of **angiosperm** with only one **seed leaf** (cotyledon) within the seed. *Monocotyledons generally have narrow leaves, and their flower parts (petals, sepals, carpels, etc.) are usually in threes or multiples of three. Monocotyledons include crop plants (oats, wheat), grasses, tulips, daffodils, and lilies.*

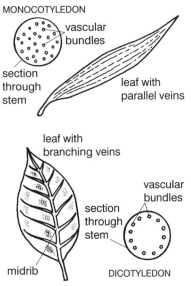

dicotyledons (abbreviation: **dicots**) are a class of **angiosperm** with two **seed leaves** (cotyledons). *Dicotyledons generally have broad leaves, and their flower parts are in fours or fives or multiples of these. There is a wide variety of dicotyledons which can be subdivided into **herbaceous plants**, **deciduous trees**, and shrubs.*

annuals are plants that complete their life cycle in one year, during which they germinate, flower, produce seeds, and die. *Examples are sunflower and marigold plants.*

biennials are plants that require two growing seasons to complete their life cycle. *In the first year food reserves are built up which are used in the second year for producing flowers and seeds. Examples are wallflower, foxglove, and carrot plants.*

perennials are plants that live for a number of years. *Perennial plants may be **herbaceous** (non-woody) or woody (shrubs and trees).*

herbaceous plants are plants which die back above ground level during the winter but have organs for survival beneath the soil. *These organs are called 'perennating organs'. They include **bulbs**, **corms**, **rhizomes**, and **tubers**.*

deciduous trees are broad-leaved plants which shed their leaves at the end of the growing season to prevent dehydration. *Deciduous trees include all the woodland trees like ash, oak, and beech. The wood of these trees is hardwood, which is used for strong or decorative woodwork.*

classification (invertebrate animals)

animal kingdom (or **Animalia**) The animal kingdom includes most multicellular organisms that cannot photosynthesise. *The animal kingdom can be split into two main groups called **vertebrates** and **invertebrates**. Most animals have certain characteristics:*
- **Heterotrophic nutrition**.
- **Movement**, to be capable of searching for their food.
- Developed **nervous system** and sense organs, to be able to respond quickly to stimuli.
- Their **cells** do not have cell walls.

heterotrophic nutrition (or **holozoic nutrition**) is a type of nutrition in which all the energy is obtained from the tissue of other organisms (plant or animal). *Animals are incapable of synthesising food from inorganic material.*

invertebrate animals are animals that do not have a **vertebral column** or spine.

cnidarians (or **coelenterates**) (phylum **Cnidaria**) are aquatic **invertebrates** which have a body made up of two layers of cells, with only one body opening which acts as both mouth and anus. *This opening is usually surrounded with tentacles. This phylum includes hydra, jellyfish, corals, and sea anemones.*

APPROXIMATE NUMBERS OF SPECIES	
protozoans	30 000
cnidarians	10 000
platyhelminths	25 000
annelids	14 000
arachnids	60 000
crustaceans	39 000
myriapods	13 000
insects	1 000 000 +
molluscs	100 000
echinoderms	6 000
(vertebrates	46 000)

platyhelminths (phylum **Platyhelminthes**) are **invertebrates** which are flatworms with unsegmented bodies. *They include planarians, flukes, and tapeworms. Most flatworms are hermaphrodite.*

hermaphrodite (or **bisexual**) refers to an organism which contains both male and female reproductive organs. *The term is applied to animals such as worms which contain both male and female organs.*

annelids (phylum **Annelida**) are **invertebrates** which are segmented worms with round bodies, such as earthworms and leeches. *Each segment is internally separated from the next and has stiff bristles (chaetae) which are used as sense organs.*

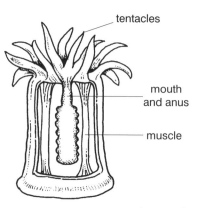

tentacles

mouth and anus

muscle

A SEA ANEMONE (CNIDARIAN)

echinoderms (phylum **Echinodermata**) are marine **invertebrates** which have spiny skins and 'sucker feet'. *These feet have water-filled canals which provide hydraulic power for movement, feeding, and respiration. This phylum includes starfish, sea urchins, sea cucumbers, and brittle stars.*

molluscs (phylum **Mollusca**) are **invertebrates** which normally have a muscular foot for movement and a soft-bodied hump or mantle which is often protected by a shell. *Members of the phylum include gastropods (snails, slugs), bivalves (mussels, oysters) and cephalopods (squids, octopuses). Cephalopods are the most advanced group of molluscs, with excellent vision and a well-developed brain.*

arthropods (phylum **Arthropoda**) are the largest and most successful phylum of **invertebrates** and have segmented bodies, jointed legs, and a hard exoskeleton.

crustaceans (class **Crustacea**) are **arthropods** which have two pairs of antennae and more than four pairs of legs. *Examples include woodlice, shrimps, crabs, and lobsters.*

arachnids (class **Arachnida**) are **arthropods** which have four pairs of legs and no antennae. *Examples include spiders, ticks, and scorpions.*

insects (class **Insecta**) are **arthropods** which have three parts to the body, three pairs of legs, and two pairs of wings (though one or both may be reduced in size). *Over 70% of all animals are insects, including ants, beetles, butterflies, fleas, bees, etc. Their* **exoskeleton** *prevents loss of water from the body, so they can live in very dry places. Many insects undergo* **metamorphosis.**

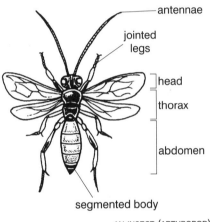

AN INSECT (ARTHROPOD)

myriapods (class: **Myriapoda**) are **arthropods** with one pair of antennae and an elongated body made up of numerous segments. *Examples are centipedes, which have one pair of legs per body segment, and millipedes, which have two pairs of legs per body segment.*

larva A larva is the juvenile stage in the life cycle of many **arthropods** such as a nymph, caterpillar, grub, or maggot. *It hatches from the egg, and is usually unlike the adult and incapable of reproduction.*

metamorphosis is the transformation that occurs in the life cycle of many **arthropods** from the egg through the larval and pupal stages to the adult form (**imago**). *It also occurs in amphibians which turn from tadpoles into adults.*

pupa The pupa is the stage of the life cycle during which the **larva** is transformed into the adult, and movement and feeding cease. *A chrysalis is the pupa formed by a caterpillar changing into a butterfly.*

imago is the sexually mature adult form of the **arthropod**. *Its appearance is often very different from the larval or pupal stages.*

classification (vertebrate animals)

chordates (phylum **Chordata**) are animals which at some stage during their development have a flexible skeletal rod or notocord running along the length of the body. *The most familiar chordates have a **vertebral column** or backbone and are called **vertebrates**.*

vertebrate animals (or **craniates**) belong to a major subphylum of **Chordata** which contains all those animals with a vertebral column. *Vertebrates are characterised by having a flexible **endoskeleton** made of bone and cartilage. They also have a complex **nervous system** and a well-developed brain. There are five classes of vertebrate: **fishes**, **amphibians**, **reptiles**, **birds**, and **mammals**.*

cold-blooded animals (or **ectotherms** or **poikilotherms**) include all **invertebrates**, **fish**, **amphibians** and **reptiles**, whose body temperature is dependent upon their environment. *Such animals become sluggish in cold weather as the low temperature slows down their metabolic rate.*

warm-blooded animals (or **endotherms** or **homoiotherms**) include all **birds** and **mammals** which can generate and maintain their body temperature no matter what the temperature of their environment. *Such animals normally have feathers or fur (hair) to help to keep their body temperature constant (36° – 38°C in mammals and 38° – 40°C in birds). High internal body temperature allows fast action of muscles and nerves. As a result such animals can be highly active even in cold climates. Some mammals (like bats) and birds (like owls) are active at night (nocturnal).*

fishes (class **Pisces**) are **vertebrates** which are aquatic and cold-blooded and have skin covered with scales. *Most fishes have gills for breathing and fins for movement. They can be divided into two main subclasses: **bony fishes** and **cartilaginous fishes**.*

bony fishes (subclass **Osteichthyes**) are marine and freshwater fishes with a bony skeleton and a swim bladder which allows them to suspend themselves in water at any depth by letting air in or out of the bladder. *Most bony fishes have a flap of skin (operculum) covering their gill slits. Examples include perch, cod, and salmon.*

cartilaginous fishes (subclass **Chondrichthyes**) are marine fishes which have a skeleton made of soft bone called **cartilage**. *Unlike bony fishes they do not have a swim bladder, and therefore avoid sinking only by constant swimming. They are mainly carnivores. Examples include sharks, rays, and skates.*

amphibians (class **Amphibia**) are cold-blooded **vertebrates** which are semi-aquatic: the female always returns to the water to lay her eggs. *Amphibians usually undergo **metamorphosis**. The larval form (tadpole) lives in water and has gills. The adult form normally lives on land and has lungs. Amphibians were the first vertebrates to occupy land (about 370 million years ago). The class of amphibians includes frogs, toads, newts, and salamanders.*

reptiles (class **Reptilia**) are cold-blooded **vertebrates** which lay soft-shelled eggs on land. *Reptiles were the first class of vertebrate to live entirely on dry land. Their skin is covered by horny scales to prevent water loss. As they live on land they breathe with lungs. Fertilisation is internal and does not require water to transfer the sperm to the egg. The class of reptiles includes snakes, lizards, crocodiles, turtles and tortoises, and the extinct dinosaurs.*

birds (class **Aves**) are warm-blooded **vertebrates** with feathers, wings and a beak. *They evolved from reptiles around 150 million years ago. Their skin is dry and has no sweat glands. Cooling to maintain a constant body temperature is achieved by panting. Fertilisation is internal and birds lay eggs with a hard shell. Although all birds have wings, some (e.g. emu, ostrich, penguin) have wings that are not large enough to support their body weight, so they cannot fly.*

mammals (class **Mammalia**) are warm-blooded **vertebrates** whose skin is covered with hair and has sweat glands. *They normally give birth to live young which are nourished in a womb with a placenta. All female mammals have mammary glands to suckle their young. Other distinguishing characteristics are the presence of a **diaphragm**, ear **ossicles** (middle ear), and different types of **teeth** (incisors, canines, and molars). Mammals evolved from carnivorous reptiles about 225 million years ago.*

marsupials are a subclass of **mammals** which raise their young in a pouch. *Examples are opossums, koalas, kangaroos, and wallabies.*

monotremes are a small subclass of **mammals** which lay eggs. *Examples are the duck-billed platypus and the spiny anteater.*

primates are the most advanced order of **mammal** with an enlarged forebrain and well developed cerebral activities, e.g. intelligence, highly developed senses (especially ears and eyes), etc. *Primates undergo a long period of growth, development, and learning, with a strong parental influence. All primates have hands and feet with fingernails, and an opposable thumb and forefinger for grasping. This order includes monkeys, apes, and humans.*

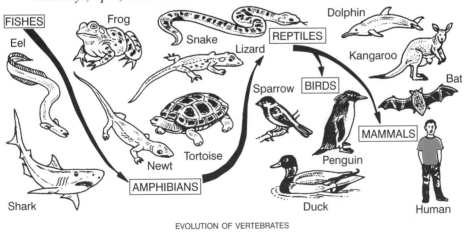

EVOLUTION OF VERTEBRATES

communication is the sending of information from one place to another.

telecommunications involves sending information over long distances either by wires or by electromagnetic radiation (radio, light, etc.).
- The message must be sent by a **transmitter**.
- The message must be placed on a **carrier**.
- The message must be detected by a **receiver**.

transmitter (radio transmitter) A transmitter is a device which converts electrical impulses into modulated **radio waves**.

radio waves are **electromagnetic waves** with frequencies between 3 kHz and 300 GHz. *Radio waves are chosen as the **carrier wave** because they travel fast (speed of light) and are easily reflected and diffracted. Radio waves can be produced by making electrons (electricity) oscillate in an **aerial**.*

aerial (or **antenna**) An aerial is the part of a radio system which is used to transmit or receive radio waves. *Transmitting aerials can be tall masts. Microwave aerials for receiving TV signals are usually dish-shaped.*

carrier wave A carrier wave is a continuously transmitted **radio wave**. *It carries a signal which is combined with the carrier wave by **amplitude modulation** or **frequency modulation**.*

amplitude modulation (abbreviation: **AM**) is a form of radio transmission in which the sound being broadcast is conveyed by variations in the **amplitude** of the radio **carrier wave**. *It is used on long and medium wavebands.*

frequency modulation (abbreviation: **FM**) is a form of radio transmission in which the sound being broadcast is conveyed by variations in the **frequency** of the radio **carrier wave**. *It is used on shorter wavebands.*

receiver (radio receiver) A receiver is a device which converts modulated **radio waves** into electrical impulses which can then be changed into sound.

transducer A transducer is a device which converts an electrical signal into a sound signal (or other energy form) or vice versa. *The transducer in the receiver circuit is usually a loudspeaker or earpiece.*

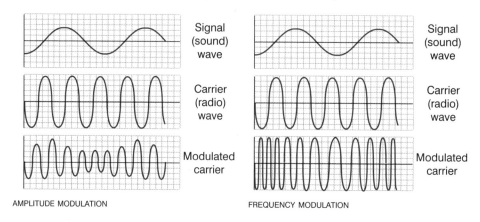

Signal (sound) wave

Carrier (radio) wave

Modulated carrier

AMPLITUDE MODULATION

Signal (sound) wave

Carrier (radio) wave

Modulated carrier

FREQUENCY MODULATION

attenuation is the gradual decrease in strength of a wave as it loses energy when passing through a medium. *This is due to absorption and scattering, and results in a gradual decrease in the amplitude of the wave as its energy is converted to heat energy.*

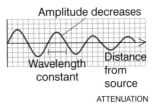

Amplitude decreases

Wavelength constant

Distance from source

ATTENUATION

polarisation is the effect of an **electromagnetic wave** oscillating in one plane only. *Broadcast radio waves are often polarised, so an aerial must be vertically aligned with their electric field to receive them.*

digital signals are signals which do not vary continuously, but normally consist of two voltage levels which can be represented as 0 (no voltage) and 1 (high voltage). *They can be boosted to compensate for loss of energy due to* **attenuation** *without any loss in the quality of the signal. Also they can be compressed ('squashed up') so that many more signals can be carried. Digital television began broadcasting in the UK in 1998.*

analogue signals are signals (electrical impulses) which have a continuous variation of voltage with time. *An example would be the voltage output from a microphone. When analogue signals are amplified, any background noise associated with the signal is also amplified, which causes signal distortion.*

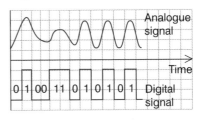

Analogue signal

Time

0 1 00 11 0 1 0 1 0 1 Digital signal

ANALOGUE AND DIGITAL

artificial satellites are man-made satellites that orbit the Earth, Moon, Sun or a planet. *Communication satellites are artificial satellites.*

communication satellites are small, unmanned spacecraft in orbit above the Earth which are used for radio, television, and telephone transmissions. *They use* **microwave transmission**, *as this is of higher energy (frequency) than normal radio waves.*

geostationary satellites are **communication satellites** that orbit anticlockwise high above the equator at a speed to match the rate of rotation of the Earth on its axis. *They therefore make one complete orbit every 24 hours and stay in the same positions above the Earth. They are used for television transmission.*

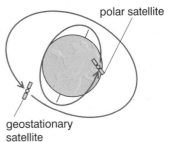

polar satellite

geostationary satellite

TYPES OF SATELLITE

polar satellites are **communication satellites** that orbit over the North and South Poles. *The size of their orbit is much less than the geostationary satellite. During the orbit they pass over all parts of the Earth's surface every few days. Such satellites are therefore useful for surveillance or for weather forecasting.*

conservation is the protection of the **environment** from the effects of **pollution** and human activity, and the sensible use of the Earth's **natural resources**. *A rapidly increasing human population has led to increased demands for food, fuel, wood, minerals and other natural resources. This must be balanced out by alternative food sources and renewable energy sources.*

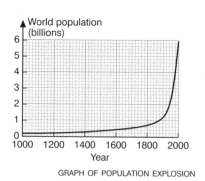

GRAPH OF POPULATION EXPLOSION

natural resources can be categorised into two types: **renewable** and **non-renewable resources**.

renewable resources include plant and animal products such as food, crops, timber and wood for fuel, and energy sources such as wind power and solar power. *It is important that the harvesting rate of such resources does not exceed their replacement rate, so allowing natural restocking.*

non-renewable resources include **minerals** and energy sources such as **fossil fuels** (coal, oil, and natural gas). *Once such resources are used up they cannot be replaced. Fossil fuels will last longer if more use is made of renewable energy sources, which also produce less pollution, or nuclear energy.*

energy conservation The use of fuels is made more efficient, so that they last longer, by reducing the loss of energy as waste heat. *It helps to use energy-efficient light bulbs, insulated buildings and economical car engines.*

deforestation is the destruction of forests by the excessive cutting down of trees for timber, paper, fuel, etc., or by the action of **acid rain** or soil erosion. *When trees have been cut down, the soil is easily washed away as there are no roots to hold it in place. Trees help to convert carbon dioxide from the atmosphere into oxygen by photosynthesis. Deforestation on a large scale (e.g. in the Amazon region of South America) creates an imbalance in this process. Levels of carbon dioxide in the atmosphere increase, contributing to the greenhouse effect.*

reforestation is the replanting of trees on land where forests used to be.

afforestation is the planting of trees to create new forests.

monoculture is a form of agriculture in which one single crop is grown continuously over very large areas. *This creates a conservation problem as natural habitats such as hedges and trees are destroyed. Large amounts of fertilisers and pesticides have to be used, which can cause pollution.*

pesticides are chemicals used in agriculture to kill pests (insects, worms) that damage crops. *These chemicals can build up in animals that feed on the pests. Birds and mammals (even humans) can be killed if high enough concentrations of pesticides such as DDT build up inside the body, often in fatty tissue.*

biological pest control This involves the use of organisms which are themselves harmless to crops but which feed on the pests. *It reduces the need for chemical pesticides and so helps to cut down on pollution. Two common crop pests are mealy bugs and aphids. Special types of ladybird can be used to attack mealy bugs and a certain species of midge larva will eat aphids.*

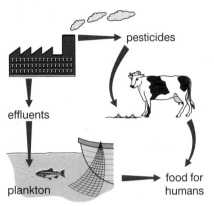

TRANSPORT OF POLLUTANTS

food production The efficiency of food production can be improved by reducing the number of stages in particular food chains, so that less energy is lost. *Normally in a food chain around 90% of biomass is lost between feeding (trophic) levels. This can be reduced by limiting the activity of animals, e.g. raising hens in battery cages. Alternative new sources of food like mycoprotein can be made from microorganisms. Such food is high in protein and can be grown quickly and cheaply. It is often used as animal feed.*

overfishing is a conservation problem caused by catching most of the fish in their feeding grounds, which does not leave enough fish to reproduce and replenish stocks. *Even small fish which are thrown back often die. Some whale (blue whale) and herring stocks around Britain have suffered from overfishing. Countries fishing in international waters now have to abide by their 'fish quotas'. Fish farms for salmon, trout, eels, etc., help to increase stocks by ensuring high levels of reproduction and quick growth.*

extinction is the complete and irreversible disappearance of all living members of a species or group of organisms from the Earth. *Examples are the dinosaurs, mammoths, and the dodo of Mauritius. Extinction of species can be caused by destruction of habitat (e.g. deforestation), by hunting (e.g. overfishing), or by disease.*

recycling is the reprocessing of used materials to save money and natural resources, after sorting waste material and rubbish into components like metal (tin cans), glass (bottles) and paper (cardboard, newspaper).

green-belt areas These are areas around towns and cities where housing or industrial development are restricted. *Such areas prevent 'urban sprawl' or 'ribbon development' along roads. Priority is given to the development of derelict sites within towns or cities.*

nature reserves These protect endangered animals and their habitats. *National parks and SSSIs (Sites of Special Scientific Interest) help to educate the public of the importance of conservation. Stricter laws control pollution and protect endangered species, e.g. banning hunting for skins, ivory, folk medicines, etc.*

corrosion (oxidation and reduction)

corrosion is a **chemical reaction** between a metal and the gases in the air. *Typically the metal reacts with oxygen to form an oxide layer on its surface (oxidation). Often this weakens the metal (as with iron), but sometimes this oxide coating forms a protective coat against further corrosion (as with aluminium).*

prevention of corrosion has to stop the oxidation of the metal, so it must either prevent oxygen from reaching the metal, or prevent the metal from losing electrons. *Coating metals (machine parts, tools, etc.) with grease or oil prevents air (oxygen) from reaching the metal. Painting metal objects or coating metal objects with plastic (garden chairs, dish racks, etc.) has the same effect and so helps to prevent corrosion.*

rusting is the corrosion of iron or steel to form hydrated iron(III) oxide $Fe_2O_3.xH_2O$. *For rusting to occur, both air (oxygen) and water must be present. Rusting is a redox reaction.*

oxidation is a **chemical reaction** involving the gain of oxygen (or the loss of hydrogen). *Alternatively oxidation can be regarded as a process which involves the loss of electrons by a substance. Many oxidation reactions are very useful, e.g. combustion to produce heat and light, and the process by which energy is released when foodstuffs and oxygen combine (respiration).*

> An oxidation reaction: burning of methane
> $$CH_4 + 2O_2 \rightarrow CO_2 + 2H_2O$$
> The carbon has gained oxygen and lost hydrogen.

reduction is a **chemical reaction** involving the loss of oxygen (or the gain of hydrogen). *Alternatively reduction can be regarded as a process which involves the gain of electrons by a substance.*

> A reduction reaction: extraction of iron
> $$Fe_2O_3 + 3CO \rightarrow 2Fe + 3CO_2$$
> The iron has lost oxygen and gained electrons.

redox reaction A redox reaction is a chemical reaction involving simultaneous **oxidation** and **reduction**. *The two processes always occur together, as when electrons are lost from a substance (oxidation) they must be gained by another substance (reduction).*

oxidising agent (or **oxidant** or **electron acceptor**) An oxidising agent is a substance which helps **oxidation** to occur. *It provides oxygen and/or accepts electrons. Examples of strong oxidising agents are potassium manganate (VII) $KMnO_4$ (acidified) and potassium dichromate $K_2Cr_2O_7$ (acidified).*

reducing agent (or **reductant** or **electron donor**) A reducing agent is a substance which helps **reduction** to occur. *It removes oxygen and/or donates electrons. Examples of good reducing agents are carbon, hydrogen, and carbon monoxide.*

galvanising is a method of protecting a metal (e.g. iron or steel) from corrosion by covering it with a thin layer of **zinc** through dipping or

electroplating. *Corrugated roofing sheets or sheets for dustbins made of mild steel are often galvanised by dipping them in molten zinc. The galvanised iron remains protected even if the surface is scratched because of sacrificial protection.*

tin plating is **electroplating** a thin layer of tin metal on both sides of a sheet of mild steel to make tinplate. *Tinplate is used in the canning industry for 'tin cans' for fruit, vegetables, baked beans, etc., because tin is unreactive and non-toxic.*

chromium plating is **electroplating** a thin layer of chromium metal onto steel to give it a shiny protective coating. *Chromium plating is used on bicycle handlebars, car bumpers, etc.*

stainless steel is an **alloy** of iron 74%, chromium 18%, and nickel 8%, which is resistant to corrosion. *Stainless steel is commonly used in making cutlery, chemical plant, and surgical instruments.*

sacrificial protection is a method of protecting a steel structure (bridge, underground pipe, etc.) by fastening to it a more reactive metal such as magnesium or zinc. *The protecting metal is more reactive, and in a damp or wet environment loses electrons in preference to the iron. It is therefore used up and periodically has to be replaced. However, the iron does not rust, as it does not release electrons and is not oxidised.*

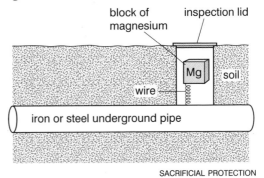

SACRIFICIAL PROTECTION

anodising is a method of coating objects made of aluminium with a protective oxide coating by **electrolysis**. *Aluminium is more reactive than iron and soon forms a thin oxide coating (10^{-6} cm). However, unlike iron oxide (rust) it does not flake off but acts as a protecting film to prevent further corrosion.*

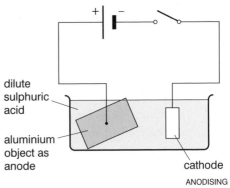

ANODISING

*Anodising makes this oxide layer thicker (10^{-3} cm) by making the aluminium the **anode** of a cell in which dilute sulphuric acid is electrolysed. Oxygen forms at the anode and reacts with the aluminium to make a thicker, more protective oxide coating.*

density (symbol: ρ) The density of a material is its mass per unit volume. *For an object of mass m and volume V, its density ρ is given as m/V. Its units are $kg\,m^{-3}$ or $g\,cm^{-3}$. If one substance has a higher density than another, then the same mass of each substance will have a different volume, and the same volume of each substance will have a different mass.*

$$density = \frac{mass}{volume}$$

relative density (or **specific gravity**) is the density of a substance compared with the density of water.

$$relative\ density = \frac{density\ of\ substance}{density\ of\ water} = \frac{mass\ of\ substance}{mass\ of\ same\ volume\ of\ water}$$

Relative densities have no units but indicate how much more or less dense a substance is than water. They can be calculated by dividing the mass of any volume of a substance by the mass of an equal volume of water.

density bottle A density bottle is a container which holds a precisely measured volume of liquid (at constant temperature). *To measure the relative density of a liquid, completely fill the density bottle with the liquid and measure its mass. Then fill the density bottle with water and find its mass. Divide the first mass by the second mass to find the relative density.*

Substance	Density kg m^{-3}	Density g cm^{-3}	Relative density
air	1.3	0.0013	0.0013
cork	250	0.25	0.25
wood (beech)	750	0.75	0.75
petrol	800	0.8	0.8
ice (at 0°C)	920	0.92	0.92
water	1000	1	1
aluminium	2700	2.7	2.7
stainless steel	7800	7.8	7.8
copper	8900	8.9	8.9
lead	11 400	11.4	11.4
gold	19 300	19.3	19.3

eureka can (or **displacement can**) A eureka can is used to find the volume of an irregularly-shaped object. *The object is completely submerged in the eureka can, which is first filled with water. The volume of the water displaced is equal to the volume of the object. If its mass is known its density can be calculated.*

measure volume of water displaced

object displaces water

EUREKA CAN

Archimedes' principle states that when a body is partially or totally immersed in a **fluid** there is an **upthrust** equal to the weight of the fluid displaced. *It was named after the Greek mathematician Archimedes (287–212 BCE). The word 'fluid' is used because the principle applies to both liquids and gases. Objects surrounded in air (such as balloons) experience upthrust as well as objects immersed in liquids. A hot air balloon rises up because of the upthrust from the surrounding denser cold air.*

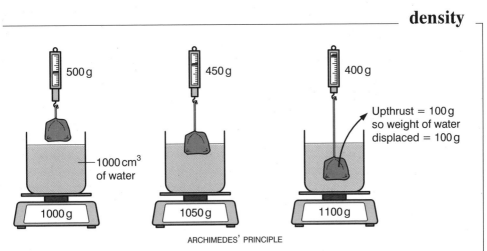

ARCHIMEDES' PRINCIPLE

upthrust is the upward **force** on an object which is immersed in a fluid. *If the upthrust is equal or greater to the **weight** of the object, then the object will float in the fluid.*

buoyancy is the result of the **upthrust** on a body which is floating or suspended in a fluid.

principle of flotation states that when a body floats its **weight** is equal to the **upthrust** on it. *As the upthrust is equal to the weight of displaced fluid (**Archimedes' principle**) we can restate the principle as 'a floating body displaces its own weight of fluid'.*

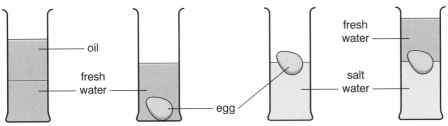

A less dense substance floats on or in a more dense substance.

floating or sinking An object will only float if its **density** is the same or less than the density of the surrounding fluid. *If an object has a higher density than the surrounding fluid, it sinks.*

hydrometer A hydrometer is an instrument for measuring the **density** or **relative density** of a liquid. *It often consists of a calibrated hollow tube which is weighted at the bottom so that it floats in the liquid. The less dense the liquid, the greater the volume which must be displaced for the upthrust to equal the weight and so the more the hydrometer is submerged.*

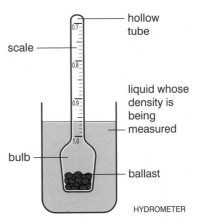

HYDROMETER

digestion is the breaking apart of ingested food into chemically simpler forms that can be easily absorbed and assimilated into the body.

extracellular digestion occurs in most animals, and takes place in the **alimentary canal**, outside the cells where respiration takes place.

intracellular digestion occurs inside the cell in **protozoans** and other single cells (such as **phagocytes**).

mechanical digestion is the physical process of breaking large pieces of food into small pieces using the teeth and the churning movements of the alimentary canal. *Smaller pieces of food provide a greater surface area for* **chemical digestion**.

mastication is the process of chewing food and involves the movement of the jaws and teeth. *Mastication is part of mechanical digestion.*

bolus A bolus is a ball of chewed food bound together with **saliva** and small enough to pass through the oesophagus.

chemical digestion is the chemical process of breaking large insoluble food molecules into smaller soluble molecules. *Enzymes are normally involved in this process.*

ingestion (or **feeding**) is the taking of food into the organism (usually through the mouth) for subsequent digestion.

absorption is the passing of soluble food molecules through the walls of the intestines and stomach into the blood stream. *Absorption normally occurs through* **villi**.

villus (plural: **villi**) A villus is one of the finger-like projections which line the wall of the small intestine. *Each villus has a large surface area to increase absorption. Inside the villus is a network of blood capillaries for absorption of soluble food material.*

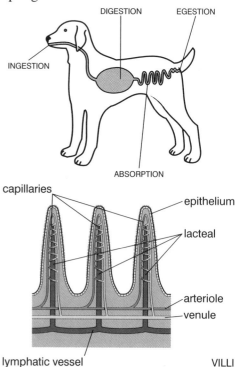

microvillus A microvillus is one of a number of minute finger-like projections which increase the surface area of a **villus**.

epithelium (or **epithelial tissue**) Epithelium is **tissue** which lines cavities or tubes in the body such as the intestines, bladder, and lungs. *Some epithelial tissue is permeable, and some is glandular, containing* **goblet cells** *which secrete* **mucus**.

lacteals are minute blind-ended **lymph** vessels found inside each **villus** in the small intestine. *Digested fats are absorbed into these lacteals.*

chyle is a milky fluid that forms from the absorption of fats into the **lacteals** of the small intestine.

enzymes are catalysts in biochemical reactions. *Certain enzymes help in the chemical breakdown of food and are called digestive enzymes. Most enzymes are **proteins**.*

Type of enzyme	Where it is made	Where it works	Which reaction it speeds up
Carbohydrase e.g. **amylase**	salivary glands pancreas intestine	mouth small intestine small intestine	breaks down **starch** into **sugars**
Protease e.g. **pepsin**, **trypsin**	stomach pancreas intestine wall	stomach small intestine small intestine	breaks down **protein** to **amino acids**
Lipase	pancreas small intestine	small intestine small intestine	breaks down **fat** to **fatty acids** and glycerol

TYPES OF DIGESTIVE ENZYME

amylases are a group of closely related **enzymes** that break down starch, glycogen, and other polysaccharides.

ptyalin (or **salivary amylase**) is an amylase enzyme which is found in saliva and begins the digestion of starch in the mouth. *Digestion is completed in the pancreas by pancreatic amylase.*

proteases are **enzymes** which break down **proteins** into polypeptides and amino acids.

pepsin is a **protease** secreted into the **stomach**. *It is secreted as the inactive form pepsinogen, which is converted to pepsin by the hydrochloric acid in the stomach.*

trypsin is a **protease** secreted into the **duodenum** in the **pancreatic juice**. *It is secreted as the inactive form trypsinogen, which is converted to trypsin by another enzyme.*

lipase is an **enzyme** in intestinal juice which breaks down **fats** into fatty acids and glycerol.

pancreatic juice is a liquid secreted by the **pancreas** which contains **enzymes** such as trypsin, amylase and lipase.

assimilation is the use of absorbed food molecules in the processes of growth, tissue repair, and reproduction.

egestion is the removal of food such as plant fibre which cannot be digested or absorbed. *It takes place through the **anus**. The material forms a semi-solid mass called **faeces**. Egestion must not be confused with **excretion**.*

faeces is undigested food that is eliminated through the anus.

defecation is the removal of **faeces** through the anus.

alimentary canal (or **digestive tract** or **gastrointestinal tract** or **gut**) The alimentary canal is a tube in the body of animals in which food is moved along by the action of **peristalsis**. *The muscular tissue of the canal squeezes the food along the canal, and the glandular tissue secretes the enzymes for chemical digestion. In most animals the canal has two openings, the mouth (for **ingestion**) and anus (for **egestion**). In between, the processes of **digestion** and **absorption** take place.*

peristalsis is the wave-like muscular contractions along the oesophagus and intestines which help to move the food through the alimentary canal. *Movement of food is helped by the lubricating action of **saliva** (from the salivary glands) and mucus (from the goblet cells in the intestine wall).*

saliva is a watery fluid secreted from the salivary glands in the mouth. *Saliva contains mucus and an **amylase enzyme (ptyalin)** which breaks down starch into maltose.*

epiglottis The epiglottis is a valve-like flap of **cartilage** which closes off the trachea or windpipe when food is being swallowed.

oesophagus (or **gullet**) The oesophagus is the section of the alimentary canal between the mouth and the stomach.

cardiac sphincter The cardiac sphincter is a muscular ring between the oesophagus and the stomach. *It relaxes to let food through. Sometimes acid escapes from the stomach into the oesophagus if this sphincter muscle is weak. The result is heartburn, which is a burning pain in the chest.*

stomach The stomach is a large muscular sac where **gastric juice** is secreted to begin digestion. *Hydrochloric acid is also secreted in the stomach, as the enzyme **pepsin** (which begins the digestion of protein) prefers acid conditions.*

gastric juice is an acidic mixture of digestive **enzymes** secreted into the **stomach** to begin the first stage of digestion.

palate, tongue, mouth, epiglottis, salivary glands, oesophagus, diaphragm, liver, bile duct, gall bladder, stomach, duodenum, pancreas, caecum, ileum, appendix, colon, anus, rectum

HUMAN ALIMENTARY CANAL

chyme is the semi-digested food that passes from the **stomach** into the small intestine.

pyloric sphincter The pyloric sphincter is a muscular ring between the stomach and the small intestine. *When it relaxes, it lets the semi-digested food (chyme) through to the duodenum.*

small intestine The small intestine is the portion of the alimentary canal between the stomach and the large intestine. *It consists of two sections called the duodenum and ileum.*

duodenum is the first section of the **small intestine** and is the main site for the **digestion** of food.

ileum is the final section of the **small intestine** and is the main site of **absorption** of food through the **villi**.

gall bladder The gall bladder is a small storage organ for the **bile** which is produced by the liver.

bile is a greenish-yellow fluid produced by the **liver** and stored in the **gall bladder**. *The bile is released along the bile duct into the duodenum. It is alkaline and so neutralises stomach acid. Bile is not an enzyme but breaks up (emulsifies) fat droplets so that enzymes like lipase can work better.*

pancreas The pancreas is a gland (in vertebrate animals) which secretes **pancreatic juice** into the duodenum. *The pancreas also has specialised groups of cells called the islets of Langerhans which produce hormones.*

exocrine glands are glands which secrete substances through a duct (tube) directly to the organ or body surface where it is needed. *Digestive glands such as the pancreas, gastric glands, and salivary glands are exocrine glands.*

large intestine (or **bowel**) The large intestine is the portion of the alimentary canal between the small intestine and the anus. *It consists of the caecum, colon, and rectum.*

caecum The caecum is a pouch in the alimentary canal between the small intestine and colon.

appendix (or **vermiform appendix**) is an outgrowth of the **caecum** which in humans is a vestigal organ and has no function in digestion. *In herbivorous animals (such as rabbits, sheep, and cows) it is highly developed and contains a large population of bacteria which are needed to break down the cellulose in grass.*

colon The colon is the section of the large intestine between the caecum and rectum. *Its main purpose is to absorb water and minerals from undigested food, leaving a semi-solid mass called faeces.*

rectum The rectum is the last section of the alimentary canal, between the colon and the anus. *It holds the indigestable faeces prior to removal through the anus.*

anus is the terminal opening of the alimentary canal, which is used for **egestion**. *It is surrounded by a muscular ring or sphincter.*

drugs and social issues

drugs are chemical substances which affect the nervous system and change some function of the body or mind. *Drugs come in many different forms (e.g. tablets, pills, liquids, powders, or suspensions) but they all work by being absorbed into the bloodstream and carried to the brain. Here the* **central nervous system** *coordinates the drug's action at various sites around the body.*

types of drug There are four main groups of drugs which affect the body in different ways: **analgesics**, **sedatives**, **stimulants**, and **hallucinogens**.

Type	Effect	Examples	Use and abuse
Analgesic	A painkiller which numbs the part of the brain that senses pain (usually without causing unconsciousness).	aspirin, paracetamol, morphine, and heroin	Powerful painkillers such as morphine and heroin are highly addictive.
Sedative	A drug which slows down the activity of the brain and therefore has a calming effect and induces sleep.	Valium, Librium, and other tranquillisers (alcohol also has a sedative effect)	Commonly used for treatment of high blood pressure and mental anxiety. Many sedatives after prolonged use are addictive.
Stimulant	A drug which speeds up the mental activity of the brain to make a person more alert.	amphetamine ('speed'), caffeine, nicotine, ecstasy, and cocaine	Some stimulants are taken medically to relieve depression. Many stimulants are addictive (e.g. nicotine in cigarettes).
Hallucinogen or **Deliriant**	A drug which causes visual images that are sensed in the mind but which do not actually exist.	LSD, ecstasy (in high doses), and cannabis are hallucinogens; solvents of various aerosol products are deliriants	There is widespread abuse of these drugs. Some are mixed with tobacco and smoked; others are taken in tablet form.

solvent abuse is the inhaling of volatile solvents found in aerosols, paint thinners, cleaning fluids, nail-varnish removers, etc. *These solvents can make the users become intoxicated. They are addictive, and regular inhalation can cause brain damage and kidney and liver failure. 'Glue sniffing' can kill.*

alcohol abuse is the excessive consumption of alcoholic drinks, which contain the drug called **ethanol**. *This is found in beer, wine, and spirits made from fruit and grain. It is a sedative which slows down the activity of the brain. It slows your reactions and increases the likelihood of accidents. It is illegal to drive a car with more than 80 mg of alcohol per 100 ml of blood (there are proposals to reduce this to 50 mg/100 ml).*

alcoholism is addiction to alcoholic drinks. *It is caused by prolonged excessive drinking, which can cause physical damage, especially to the liver and stomach, and also mental deterioration. An alcoholic who stops drinking becomes depressed and anxious, and suffers withdrawal symptoms such as delirium and uncontrollable shaking.*

One unit is equivalent to

| One measure (single) of WHISKY | or | One glass of SHERRY | or | One glass of WINE | or | Half a pint of ordinary BEER OR LAGER |

UNITS OF ALCOHOL

smoking There is now definite evidence that smoking cigarettes can seriously damage your health by causing heart disease, high blood pressure, and lung diseases (bronchitis, emphysema, and lung cancer). *The main harmful components of cigarette smoke are the following:*
- **nicotine**, an addictive stimulant which increases blood pressure and makes your heart beat faster.
- **tar**, which contains thousands of different chemicals, some of which are carcinogenic (cancer-causing).
- **irritants**, which are chemicals that irritate the lungs and damage the lining of its passages. Irritants also reduce the sense of taste and smell and can cause ulcers in the stomach.
- **carbon monoxide**, a poisonous gas which combines with the **haemoglobin** in the blood in preference to oxygen. It therefore lowers the oxygen content of the blood and body cells. Smoking during pregnancy increases the possibility that babies will be underweight or stillborn.

passive smoking is breathing in cigarette smoke from someone else smoking. *Children are susceptible to this if their parents smoke. Passive smoking increases the risk of developing smoking-related diseases.*

AIDS (abbreviation for **acquired immune deficiency syndrome**) is caused by the **HIV virus**, which is transmitted through the exchange of body fluids, often during sexual intercourse. *This disease damages the immune system of the body by making the **lymphocytes** (white cells in the blood which fight disease) inactive. The body becomes vulnerable to infection, especially by pneumonia, brain infections, severe diarrhoea, and an unusual type of skin cancer called Karposi's sarcoma. At present AIDS cannot be cured, although there are drugs to fight the virus. Most AIDS patients die within a few years of developing the full-blown disease.*

HIV (abbreviation for **human immunodeficiency virus**) is the virus which causes **AIDS**. *This virus is found in body fluids such as semen, vaginal fluid, and blood. HIV can be spread:*
- when an infected person has sexual intercourse with another person
- when a person uses contaminated needles from syringes when abusing drugs, ear-piercing, or tattooing
- during blood transfusions if blood is passed from an infected donor
- during pregnancy or childbirth from the mother to the baby.

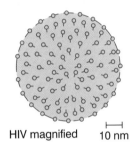

HIV magnified ⊢—⊣ 10 nm

ear The ear is the sensory organ for **hearing**, and in vertebrates also for the maintenance of **balance**.

hearing is the sense by which we detect sounds.

balance is the sense by which the brain detects the position of the head, so that it can coordinate the muscles to keep the body upright.

structure of the human ear The human ear can be divided into the **outer**, **middle** and **inner** ear. *The outer and middle ear both contain air, but the inner ear is filled with liquid.*

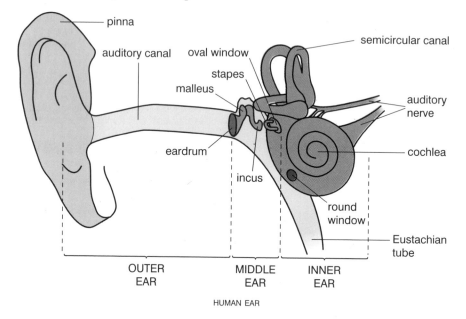

HUMAN EAR

outer ear (or **external ear**) The outer ear is the part of the ear external to the **eardrum**. *It includes the ear flap (pinna) which is responsible for directing the sound along the auditory canal to the eardrum. The lining of the auditory canal contains special sebaceous glands which secrete wax.*

eardrum (or **tympanum** or **tympanic membrane**) The eardrum is a membrane at the end of the auditory canal which transmits sound vibrations from the outer ear to the **ossicles** of the middle ear.

middle ear The middle ear is an air-filled cavity between the outer and inner ear. *It contains the ossicles.*

ossicles The ossicles are a chain of three small bones in the **middle ear** which amplify the vibrations of the **eardrum**.
- The **malleus** (or **hammer**) is the first **ossicle**, which vibrates as it touches the eardrum.
- The **incus** (or **anvil**) is the second **ossicle**, which transmits the vibration from the malleus to the stapes.
- The **stapes** (or **stirrup**) is the third **ossicle**, which causes the **oval window** to vibrate at the same frequency as the eardrum.

oval window (fenestra ovalis) The oval window is a membrane-covered opening between the **middle** and **inner ear**. *It transmits vibrations to the cochlea.*

Eustachian tube The Eustachian tube connects the middle ear to the back of the throat. *It helps to equalise the pressure on both sides of the eardrum, allowing it to vibrate. Unequal pressure causes the ears to 'pop'.*

round window (fenestra rotunda) The round window is a membrane-covered opening between the **middle** and **inner ear**. *Every time the oval window bulges inwards the round window bulges outwards. This regulates changes in pressure of the liquid inside.*

inner ear The inner ear is a structure in vertebrates which is surrounded by bone of the skull and which contains the organs of **hearing** and **balance**.

cochlea The cochlea is a spiral tube of the **inner ear** which produces nervous impulses in response to sound waves. *The vibrations of the oval window are transmitted through a fluid (perilymph) which fills the cochlea and causes the membrane of the cochlea to move up and down. Sensory hairs on the cochlea membrane produce nervous impulses which are sent to the brain along the auditory nerve. The brain interprets these impulses as sensations of sound.*

semicircular canals are the organs of balance in the **inner ear**. *These three canals are at right angles to each other and are filled with fluid (endolymph). Each has a swelling at its base called an ampulla containing tiny sensory hair cells. Movement of the fluid, caused by motion of the head, stimulates the sensory hairs to produce nervous impulses. These travel to the brain along the auditory nerve. The brain then interprets these impulses and coordinates muscle movement to keep the body upright.*

auditory nerve carries nervous impulses from the **inner ear** to the brain.

human hearing range The human ear can detect sounds with frequencies between 20 and 20 000 Hz. *The ability to hear sounds in the upper part of the frequency range decreases with age.*

audible sound is sound within the **human hearing range** (see **loudness** of sound).

deafness is an inability to hear, which can be partial or total, and either permanent or temporary. *Too much wax in the ear can cause temporary deafness. A visit to your doctor to have your ears syringed will soon restore your hearing. Permanent deafness could be due to infection or damage to the eardrum. When this happens the small bones in the middle ear do not vibrate correctly. Some people are deaf because the nerve cells in the cochlea of the inner ear are damaged and do not send any nerve impulses to the brain. As people grow older the nerve cells in the cochlea wear out and are not replaced.*

hearing aids are used by people with hearing difficulties. *They are placed close to the ear and contain a small electrical amplifier. This makes the sound louder, so that it can be picked up by fewer sensory cells.*

Earth's structure and earthquakes

Earth's structure The Earth's structure consists of distinct layers called the **crust, mantle, outer core,** and **inner core.**

Earth's crust The Earth's crust is the outer 'skin' of the Earth, consisting of large plates of rock which are floating on the mantle. *It is very thin in comparison with the overall diameter of the Earth. Its thickness varies between 5 km under the oceans to 70 km under the highest mountains. Its density is 2 to 3 g cm^{-3} and it includes **continental crust** and **oceanic crust.***

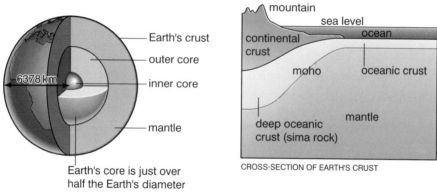

Earth's core is just over half the Earth's diameter

EARTH'S STRUCTURE

CROSS-SECTION OF EARTH'S CRUST

continental crust The continental crust rests above the oceanic crust and forms the continents. *Its main rock component is granite and it is rich in the elements <u>si</u>licon and <u>al</u>uminium (called 'sial' or low-density rocks). Much of it is very old and dates back almost to the formation of the Earth.*

oceanic crust The oceanic crust forms the bottom of the oceans and seas, and also lies deep below the less dense continental crust. *It is continually being created and remelted, so the oldest oceanic crust is only 100 million years old. Its main rock component is basalt and it is rich in the elements <u>si</u>licon and <u>ma</u>gnesium (called 'sima' or high-density rocks).*

moho (or **Mohorovičić discontinuity**) The moho is the boundary between the **Earth's crust** and the **mantle.** *It was named after the Yugoslavian geophysicist Andrya Mohorovičić (1857–1936).*

mantle The mantle is a thick layer of dense, semi-liquid rock which extends some 2900 km below the Earth's crust. *The density of this layer is 3.4 to 5.5 g cm^{-3} and the rock is rich in silicon and magnesium. Pressure and temperature increase deeper in the mantle, but even at the surface of the mantle some rocks are hot enough to be molten. **Convection currents** inside the mantle cause the plates of the Earth's crust to move (see **plate tectonics**).*

magma is hot molten rock that originates from the Earth's mantle. *Magma is extruded as lava on to the Earth's surface as a result of volcanic activity. When it cools and solidifies it forms **igneous rock.***

outer core The outer core is formed of dense liquid rock at very high temperatures, composed mainly of the dense magnetic elements nickel and iron. *Its density is 10 to 12 g cm^{-3}. The Earth's magnetic field arises* *from **convection currents** in the outer core, which generate electric currents and make the outer core act as an electromagnet.*

inner core The inner core is formed of solid rock at the centre of the Earth which is extremely dense and at very high temperature and pressure. *The density of the inner core is 12 to 18 g cm⁻³.*

The density of the inner core is 12 to 18 $g\,cm^{-3}$.

earthquakes are sudden movements of the Earth's crust caused by the plates moving against one another. *As the plates try to move relative to each other, strains build up. Eventually the tension is released, causing the ground to shake violently. The energy released travels through the Earth as a series of shock waves called **seismic waves**. Earthquakes occur mainly along the boundaries of plates in the crust.*

focus The focus is a point inside the Earth's crust where an earthquake originates.

epicentre The epicentre is the point on the Earth's surface directly above the focus of an earthquake.

seismic waves are the shock waves of an earthquake. *There are three types: **p-waves, s-waves, and l-waves.***

- **p-waves** (or **primary waves**) are **longitudinal waves** which make rock particles vibrate backwards and forwards in the same direction as the motion. They travel quickly and are the first to be detected. They can travel through solids and liquids and can therefore pass through the Earth's core.
- **s-waves** (or **secondary waves**) are **transverse waves** which make rock particles vibrate at right angles to the direction of motion. Such waves cannot travel though liquids, and therefore cannot pass through the outer core. **S-waves** travel more slowly than **p-waves** but faster than **l-waves**.
- **l-waves** (or **long waves**) are the slowest-moving shock waves, and travel only through the Earth's crust. They make the ground move, and are responsible for most of the damage that an earthquake can cause.

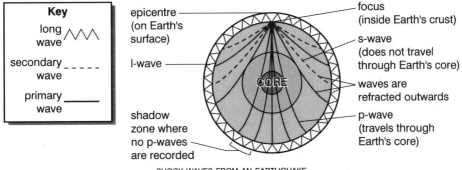

SHOCK WAVES FROM AN EARTHQUAKE

shadow zone is an area where p-waves are not recorded because they are refracted due to the increasing density of the core with depth.

Richter scale The Richter scale is a logarithmic scale of 1 to 10 used to compare the magnitude of earthquakes. *A value of 2 on the Richter scale can just be felt as a tremor. Values of 6 and above can cause damage to buildings.*

ecology is the study of the interaction between different living **organisms**, and between them and their **environment**.

environment The environment consists of all the conditions which surround an organism and in which it lives.

abiotic factors are the non-living factors which influence the environment, including **climatic factors** and **edaphic factors**.

climatic factors include sunlight, rainfall, temperature, and humidity.

edaphic factors are the chemical and physical aspects of the environment such as the oxygen content of water, the pH of soil, and the degree of air pollution.

biotic factors are factors arising from the activities of living organisms (including humans) which influence the environment. *Biotic factors include availability of food, number of predators, competition from other organisms, disease, and the impact of human activities.*

biosphere is a general term for the region of the Earth (including the air and sea) which may be inhabited by living organisms. *Life exists where energy from the sun can interact with air, water, and substances in the Earth's crust. The biosphere is the sum of the world's ecosystems.*

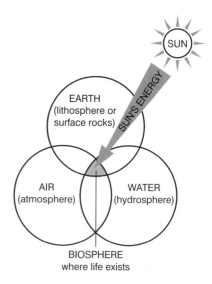

ecosystem An ecosystem is a biological community and the physical environment that is associated with it. *Examples of natural ecosystems are woodland, meadows, hedgerows, sand-dunes, seashore, marshland, pond, river, sea, etc. A greenhouse is a managed ecosystem, in which the temperature, humidity, and light can be regulated and pests controlled.*

population A population is a group of individuals of the same **species** within a community. *The size of a population can vary according to physical factors like drought, and other factors such as the **predator-to-prey ratio**.*

community A community is the total collection of living organisms (both plants and animals) living within a defined area or **habitat**. *Communities are often named after one of their dominant species or a major physical characteristic of the area.*

habitat A habitat is a place in which an organism or a **community** of organisms live. *Habitats are named after their major physical characteristic, e.g. river, forest, grassland. Sometimes large habitats (macrohabitats) have the same name as the ecosystem, but a habitat is* *just a place, whereas the ecosystem is the place and all its living organisms. Within most habitats are smaller habitats (microhabitats), e.g. rotting tree, oak tree, etc.*

studying ecosystems involves identifying living organisms within the ecosystem and estimating the abundance of each species. *Identifying living organisms can be helped by using biological keys. Estimating the abundance of a species can be made by taking samples using a **quadrat** or a **transect**. All samples must be random, and the larger your sample the more reliable your estimate of abundance. If organisms move around, you could use **mark, release, and recapture**.*

quadrat A quadrat is a square frame (normally with sides of 50 cm) which is used for ecological sampling. *The quadrat is randomly placed on the ground and the species inside are counted and recorded. To get an accurate indication of numbers of species you will need to*

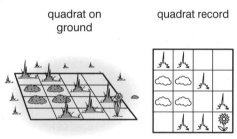

quadrat on ground

quadrat record

QUADRAT

*sample with many quadrats and average your results. A quadrat may also be used to take a sample along a **transect**.*

transect A transect is a straight line across your field or habitat along which ecological measurements are made. *You can use a long tape measure and, perhaps every 50 cm, record which plant species touch the tape. Transects are particularly useful where one kind of habitat changes into another.*

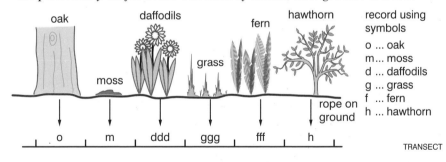

oak   daffodils   fern   hawthorn

moss   grass

rope on ground

record using symbols

o ... oak
m... moss
d ... daffodils
g ... grass
f ... fern
h ... hawthorn

o   m   ddd   ggg   fff   h

TRANSECT

mark, release, and recapture is a method used to estimate the population of organisms which move around. *First capture, let's say, 20 woodlice and mark each with a small spot of waterproof paint. Release these woodlice. After about 24 hours capture another 20 woodlice. See how many have the spot of paint. If 2 out of 20 do, then we can assume that $\frac{2}{20}$ or $\frac{1}{10}$th of the population have been recaptured, so 20 woodlice is about a $\frac{1}{10}$th of the total population. The total population is around 200 woodlice.*

Tullgren funnel A Tullgren funnel is a special funnel used to collect small animals (insects, beetles, etc.) from soil samples of leaf litter.

producers are organisms that can make their own food by **autotrophic nutrition** and are therefore considered as a source of energy. *Producers form the beginning of all **food chains**: the most important are green plants.*

consumers are organisms that feeds on others below them in a **food chain**. *All consumers feed by **heterotrophic nutrition**.*
- **primary consumers** feed on producers (plants).
- **secondary consumers** feed on primary consumers.
- **tertiary consumers** feed on secondary consumers.

herbivores are animals like cattle, deer, rabbits, and sheep which feed on plants. *These animals have grinding teeth (molars) and an alimentary canal that can digest cellulose. All herbivores are **primary consumers**.*

carnivores are animals like cats, bears, dogs, and wolves which eat meat. *These animals have well-developed canine teeth and are **predators** or carrion-eaters. All carnivores are either **secondary** or **tertiary consumers**.*

omnivores are animals which can eat both meat and plants. *Examples include humans, monkeys, and pigs.*

saprophytes (or **saprotrophs**) are organisms such as **bacteria** or **fungi** that feed on dead organic (plant or animal) matter. *Bacteria help to break down protein and fungi break down cellulose.*

decomposers are **saprophytes** that fulfil a vital role in the **ecosystem** by returning organic matter to the soil as inorganic matter, which can then be taken into plants again as mineral salts (see **natural cycles**).

detritus is organic material formed from dead and decomposing plants and animals. *Decomposers feed on detritus.*

food chain A food chain is a feeding relationship between organisms in an ecosystem. *Producers (usually green plants) begin a food chain as they are the only organisms that can make their own food. Biomass is then transferred through the food chain to **consumers** (primary to secondary to tertiary) which are usually animals.*

food web A food web is a network of interrelated **food chains** representing the complex feeding relationships of organisms in an ecosystem. *All food webs contain **producers**, **consumers**, and **decomposers**.*

trophic level (or **feeding level**) A trophic level is the position an organism occupies in a food chain. *Producers make up the first trophic level, and **consumers** make up the second and third levels.*

pyramid of numbers is the number of organisms at each **trophic level**

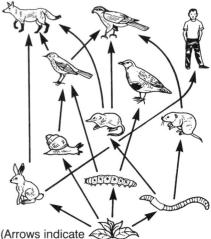

(Arrows indicate direction of biomass.)

FOOD WEB

represented pictorially as a pyramid. *The number of organisms at higher trophic levels becomes smaller. This is because at each trophic level the organisms use up most of the **biomass** they have obtained in respiration to obtain energy. This leaves less biomass (typically around 10%) to pass to higher trophic levels, so higher levels support smaller numbers of organisms.*

Trophic level 4 — T_4 fox
T_3 thrush
Trophic level 3 — T_2 caterpillar
Trophic level 2
T_1 grass

PYRAMID OF NUMBERS is the number of individuals at each trophic level

biomass is the mass of all the organisms at a particular trophic level.

Trophic level 4 — T_4 fox
T_3 thrush
Trophic level 3 — T_2 caterpillar
Trophic level 2 — T_1 grass

pyramid of biomass is the biomass of each **trophic level** represented pictorially as a pyramid. *The biomass decreases as we go up the*

PYRAMID OF BIOMASS is the total mass of individuals at each trophic level

food chain, as the number of organisms at each trophic level decreases. This pyramid is a more accurate representation of the flow of biomass through the food chain than the pyramid of numbers.

parasitism is a feeding relationship in which one living organism (the **parasite**) feeds on another living organism (its **host**) and has a harmful effect on it.

parasite A parasite is the partner in **parasitism** which benefits from the relationship, e.g. tapeworm, flea, louse, mistletoe.

host A host is the organism on which a **parasite** lives.

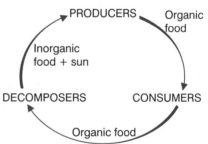

FEEDING RELATIONSHIPS

mutualism (or **symbiosis**) is a feeding relationship between two organisms from which both benefit. *An example is lichen, which consists of two organisms, a fungus and a green alga. The fungus extracts minerals from the rock or soil, and the alga photosynthesises.*

commensalism is a feeding relationship between two organisms where one benefits but the other is not harmed. *An example is a sea anemone living on the shell of a hermit crab and eating bits of food that the crab drops.*

predators are animals that hunt, kill, and eat other animals called their prey. *All predators are **carnivores**, but not all carnivores are predators.*

prey is an animal that is a source of food for a predator.

predator-to-prey ratio is the natural balance between **predator** and **prey**, which causes the **populations** of many species to remain roughly the same size over a period of time (unless there is environmental change).

electricity (statics)

static electricity (or **frictional electricity**, or **static** for short) is the accumulation of **electric charge** on an object which is a poor conductor of electricity or is insulated in some way. *It is caused by the removal of electrons from atoms by friction. Friction does not create charge: it just separates out existing charges.*

electric charge is the overall excess or deficiency of **electrons** on an object. *If there is an excess of electrons, the object has an overall negative charge. If there is a deficiency of electrons, the object has an overall positive charge.*

positive charge results when electrons are 'rubbed off' from the outermost shells of atoms. *Normally atoms have equal numbers of **protons** (positive +) and **electrons** (negative −), so the overall net charge of a material is zero. If a Perspex (acetate) rod is rubbed with a woollen cloth, some of the electrons on the rod are rubbed off on to the cloth, and the rod becomes positively charged.*

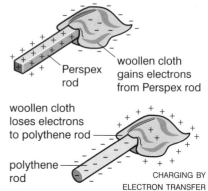

woollen cloth gains electrons from Perspex rod

Perspex rod

woollen cloth loses electrons to polythene rod

polythene rod

CHARGING BY ELECTRON TRANSFER

negative charge results when electrons are gained by being 'rubbed off' on to a material. *If a polythene rod is rubbed with a woollen cloth, some of the electrons on the cloth are rubbed on to the rod, and the rod becomes negatively charged.*

law of electrostatics The law of electrostatics states that like charges (two positive charges or two negative charges) repel one another, and unlike charges (positive and negative charges) attract each other. *The closer the charges, the greater the force between them.*

conductor A conductor is a material through which electrons can flow. *Metals are the best conductors, as their outermost electrons are loosely held and can more move freely between atoms. Carbon, water, and earth are also conductors. Conductors can become charged by static electricity, but only if they are held in insulated handles.*

insulator An insulator is a material which allows no electrons (or very few) to pass through. *In such materials the electrons are tightly held in the atoms and cannot move. Rubber, glass, air, and plastics are insulators. Insulators can become charged with static as they do not allow the electrons to flow away.*

electroscope An electroscope is an instrument for detecting small amounts of **electric charge**. *In the gold leaf electroscope the insulated metal rod becomes charged so the gold leaf is repelled away from the rod. The greater the charge, the more the gold leaf rises.*

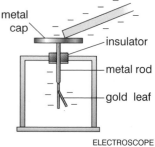

metal cap

insulator

metal rod

gold leaf

ELECTROSCOPE

induction of charge is caused by the attraction of opposite charges and the repulsion of like charges. *It is for this reason that a comb, charged by pulling it through your hair, will pick up tiny pieces of paper. It is also why dust is attracted to objects with static charge, such as TV or computer screens. Charged particles will always attract small uncharged particles by induction.*

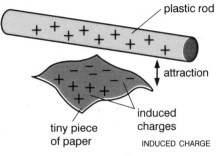

plastic rod

attraction

induced charges

tiny piece of paper

INDUCED CHARGE

point action occurs when electric charge becomes so concentrated at a sharp point that it can ionise (remove electrons from) surrounding air molecules. *These ions are then repelled by the point, creating an 'electric wind' of air molecules.*

lightning is the sudden flow of electricity from a thundercloud which has become charged by the rubbing together of air and water molecules. *As the charge builds up in the cloud so does the voltage, until suddenly a giant spark of lightning results. The **point action** of a lightning conductor can conduct this flow of electricity safely down to earth.*

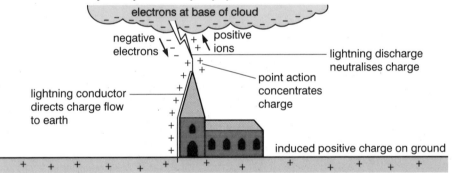

electrons at base of cloud

negative electrons

positive ions

lightning discharge neutralises charge

point action concentrates charge

lightning conductor directs charge flow to earth

induced positive charge on ground

static and safety Whenever poor conductors are rubbed together in a dry atmosphere, static can build up, causing a spark, and if a poor conductor is close to something flammable there is a risk of explosion. *Examples are the build-up of static on the plastic of petrol pipes, the chutes into grain containers (silos), and the rollers in a paper mill. It is important that such material is earthed in some way.*

uses of static electricity

- **Electrostatic smoke** and dust **precipitators** are used in chimneys to attract tiny particles of smoke and dust.
- Many photocopiers use static to form the image of a document on a charged drum. The charged areas of the drum attract graphite particles which stick to a resin coating. When paper is pressed against the heated drum, the resin and carbon particles are forced on to the paper fibres.
- Electrostatic spraying is used to apply a very even coat of charged paint droplets to an object (such as a car body) which is oppositely charged.

electricity (charge and potential)

electric charge (or **charge**) All electrons are identical and carry the same tiny quantity of charge. *The movement of electrons produces an **electric current**.*

coulomb A coulomb is the quantity of **electric charge** transported by an **electric current** of 1 amp flowing for 1 second. *The charge of a single electron is 1.602×10^{-19} coulombs. Therefore 1 coulomb is equal to a charge of about 6 million million million electrons.*

amp hour An amp hour is a quantity of **electric charge** equivalent to 1 amp flowing for 1 hour. *It is equal to 3600 **coulombs**. Car batteries are often rated in amp hours. A 40 amp hour battery will deliver 1 amp for 40 hours or 2 amps for 20 hours.*

faraday A faraday is a quantity of **electric charge** equivalent to the **Avogadro constant** (6.02×10^{23}) of electrons. *It can be defined as a **mole of electrons** and is equal to about 96 500 **coulombs**.*

electric field An electric **field** is a region within which a particle bearing an **electric charge** experiences a force. *The force is represented by 'electric field lines' which never cross each other. The intensity of the electric field is shown by the closeness of the lines. The direction of an electric field line shows the path that would be taken by a positive charge which was free to move in the field.*

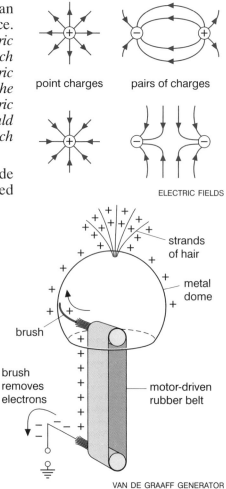

point charges   pairs of charges

ELECTRIC FIELDS

Van de Graaff generator A Van de Graaff generator is a machine used to produce electric charge from the mechanical movement of a rubber conveyor belt. *The moving belt has electrons rubbed off it, so it becomes positively charged. This positive charge is transferred to a large metal dome mounted on a hollow insulating support. The region around the charged metal dome is its **electric field**. If a tuft of hair is inserted in the charged dome, the hair stands on end. This is due to the repulsive forces of like charges between each strand. The hair has **electric potential**.*

strands of hair

metal dome

brush

brush removes electrons

motor-driven rubber belt

VAN DE GRAAFF GENERATOR

electric potential (or **electric potential energy**) is the energy associated with a charge at a particular point within an **electric field**. *It is calculated by considering the work which must be done to move a small positive charge from 'earth' to that particular point. For convenience, earth potential is given a value of zero.*

$$\text{electric potential (V)} = \frac{\text{work done (J)}}{\text{charge moved (C)}}$$

*The electric potential needed to do one joule of work on a charge of one coulomb is the equivalent to one **volt**.*

potential difference (abbreviation: **p.d.**) is the difference in potential between two charged points. *It is equal to the energy associated with the movement of a unit positive charge from one point to the other in an electric field. There is an energy change of one joule if a charge of one coulomb moves through a potential difference of one volt. Electrons always flow from a low to a high potential: that is, from the negative to the positive terminal in a battery.*

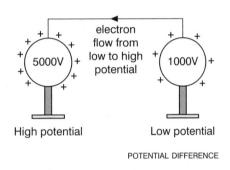

POTENTIAL DIFFERENCE

capacitor (or **electrical condenser**) A capacitor is an electrical device designed to store small quantities of electric charge. *Typically it consists of two parallel metal plates separated by an insulating material called a dielectric. This may be air, paper impregnated with oil or wax, plastic film, or ceramic. The **capacitance** of a capacitor depends on the dielectric used. The capacitance increases if the size of the plates increases, or if the distance between them becomes less. Capacitors are used in many electrical and electronic curcuits. Besides storing charge they can be used to block **direct current** while allowing **alternating current** to pass. They can also be used for time delays, as it takes time for a capacitor to charge up.*

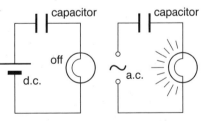

A CAPACITOR BLOCKS DIRECT CURRENT

capacitance is the ratio of the electric charge on one of the metal plates of the capacitor to the potential difference between the two metal plates. *The relationship is:*

$$\text{capacitance (F)} = \frac{\text{charge on conductor (C)}}{\text{potential difference between conductors (V)}}$$

The unit of capacitance is the farad (F) or coulomb per volt.

electricity (current and voltage)

electric current (symbol: *I*) An electric **current** is a flow of **electric charge** through a conductor. *If the conductor is a metal wire in an **electrical circuit**, then the electric charge is a flow of **electrons**. The size of the electric current is the rate of flow of electrons. Current in an electrical circuit is measured in **amperes**.*

ampere (or **amp** for short; symbol: **A**) One ampere is equal to one coulomb of electric charge passing any point in a conductor in one second. *The ampere is the SI unit of current.*

electrical circuit (or **circuit**) An electrical circuit is a continuous conducting path along which electric current can flow. *This circuit may include a variety of electrical components such as lamps, resistors, and ammeters. These components may convert the electrical energy carried by the current into other forms of energy such as heat or light. Current is not 'used up' by the electrical components, but electrical energy is transferred.*

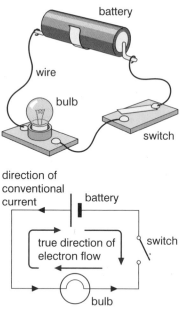

ELECTRICAL CIRCUIT

conventional current direction is from the positive terminal of the battery to the negative terminal and is shown as an arrow on the circuit diagram. *This convention was decided before it was realised that current was a flow of negatively charged electrons from the negative terminal of the battery to the positive terminal.*

battery in a circuit A battery in a circuit gives **potential energy** to the electrons which come from it. *It can only 'push out' electrons when the electrical circuit between both terminals of the battery is complete. The electrons travelling around the circuit lose all this potential energy. The energy may be given off as heat when it passes through a wire, or light as it passes through a bulb. The same number of electrons reaches the other terminal of the battery, so there is no loss in current. It is only the potential energy of the electrons that is lost.*

potential difference of battery (abbreviation: **p.d.**) The potential difference across the terminals of a battery indicates the potential energy given to each coulomb of charge 'pushed' out. *There is a p.d. of one **volt** across the battery if each **coulomb** of charge is given one **joule** of potential energy. This potential difference gradually lessens around the circuit. The sum of all the p.d.'s around the circuit is equal to the p.d. across the battery.*

electromotive force (abbreviation: **e.m.f.**) is equivalent to the **potential**

difference across the terminals of a battery when it is not supplying a current. *Like all electrical components, batteries have a resistance. This is called 'internal resistance' This causes a drop in p.d. when the circuit is complete and current flows. Electromotive force can be regarded as the 'total p.d.' including the p.d. lost across the internal resistance of the battery. Electromotive force is measured in volts.*

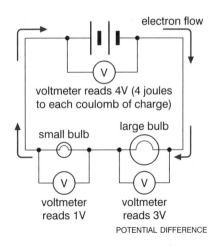

electron flow

voltmeter reads 4V (4 joules to each coulomb of charge)

small bulb

large bulb

voltmeter reads 1V

voltmeter reads 3V

POTENTIAL DIFFERENCE

voltage (symbol: **V**) is the **potential difference** or the value of this.

volt (symbol: *V*) One volt is the p.d. between two points when one coulomb of electricity passes between these points and produces one joule of work. *A volt is the SI unit of* **potential difference** *or* **electromotive force**.

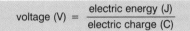

$$\text{voltage (V)} = \frac{\text{electric energy (J)}}{\text{electric charge (C)}}$$

series circuit A series circuit is formed when the components are arranged so that there is a single path for the current to take. *The current passes through each component one after another. The potential difference (voltage) across the battery is equal to the sum of all the p.d.'s across the components. A disadvantage of series circuits is that if one component stops working then no current will flow through the circuit.*

parallel circuit A parallel circuit is formed when the components are arranged so that there is more than one path for the current to take. *The current splits up and passes through each branch at the same time. The size of the current in each branch depends on the resistance of the branch. The total current is equal to the sum of all the currents in the branches. The p.d. (voltage) is the same across each parallel component. If a component in a parallel circuit breaks then current can still pass through another branch.*

ammeter An ammeter is an instrument used to measure the amount of **electric current** flowing through a particular point in an electrical circuit. *An ammeter must be connected in* **series**.

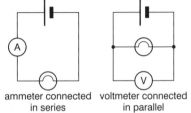

ammeter connected in series

voltmeter connected in parallel

voltmeter A voltmeter is an instrument used to measure the **potential difference** (voltage) between any two points in an electrical circuit. *A voltmeter must be connected in* **parallel** *across the component whose potential difference it is measuring.*

electricity (resistance)

Ohm's law states that the ratio of the potential difference across the ends of a metal conductor to the electric current flowing through the conductor is a constant. *This constant is the **resistance** of the conductor. The law was discovered in 1827 by the German physicist Georg Ohm. It is often expressed in an equation:*

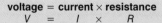

$$\text{voltage} = \text{current} \times \text{resistance}$$
$$V = I \times R$$

ohm (symbol: Ω) An ohm is the **resistance** of a conductor in which a current of one ampere flows when a potential difference of one volt is applied across its ends. *An ohm is the derived SI unit of electrical resistance.*

resistance (or **electrical resistance**) is the ability of a conductor to resist, or oppose, the flow of an electric current through it. *All of the components in an electric circuit have a certain resistance to current. This makes the electrons lose some of the electrical energy which they carry. A bulb has a very high resistance and converts electrical energy into heat and light energy. A conductor's resistance depends on the type of material it is. Resistance of a component decreases if the area of its cross-section is increased, but increases if its length increases.*

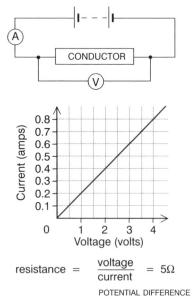

$$\text{resistance} = \frac{\text{voltage}}{\text{current}} = 5\Omega$$

POTENTIAL DIFFERENCE

flow of water analogy If you regard the amount of water flowing through a pipe as 'current' and the pressure of the water as 'voltage', then 'resistance' is any restriction to the flow of water. *Narrow pipes will have high resistance and create high pressure (high voltage). Wide pipes will have low resistance which will allow lots of water to flow (large current).*

resistor A resistor is a component of an electrical circuit that is present because of its electrical resistance. *Resistors are often included in circuits to limit the current passing through components and reduce the danger caused by overheating.*

electrical conductivity is the ability of a substance to allow the passage of an electric current. *Good **conductors** have a high electrical conductivity. **Insulators** have a low conductivity.*

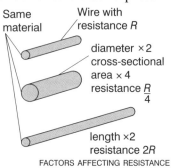

FACTORS AFFECTING RESISTANCE

superconductivity is the absence of electrical resistance at temperatures close to **absolute zero** (zero Kelvin). *Only certain metals show superconductivity, as the flow of current often heats up the metal so that it cannot reach very low temperatures.*

non-ohmic conductors are conductors which do not obey Ohm's law and do not have a constant resistance. *The higher the resistance of a conductor, the more difficult it is for the electrons to pass through. Electrical energy is therefore changed into heat energy: the temperature of the conductor rises and so does its resistance. Thin wires, such as filament bulbs, can glow white hot. A graph of current against potential difference for non-ohmic conductors is a curve.*

potential divider (or **voltage divider**) A potential divider is a chain of resistors in series (or one continuous long resistor) that can be tapped at one or more points to obtain a known fraction of the total voltage across the chain (see diagram).

resistors in series have a total combined resistance (R) which is equal to the sum of all the resistors in the circuit.

R	$=$	$R_1 + R_2 + R_3$
combined resistance		individual resistances

resistors in parallel have a total combined resistance (R) which can be calculated as follows:

total current = sum of all currents in parallel circuit

$$I = I_1 + I_2 + I_3$$

$$\frac{V}{R} = \frac{V}{R_1} + \frac{V}{R_2} + \frac{V}{R_3}$$

so $\frac{1}{R} = \frac{1}{R_1} + \frac{1}{R_2} + \frac{1}{R_3}$

The reciprocal of the combined resistance is the sum of the reciprocals of the individual resistances.

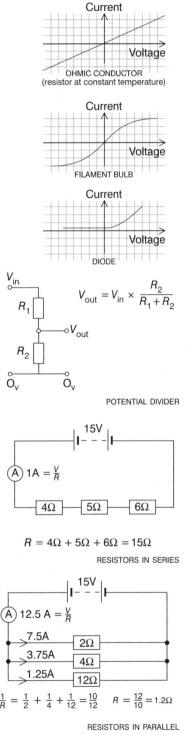

OHMIC CONDUCTOR
(resistor at constant temperature)

FILAMENT BULB

DIODE

$$V_{out} = V_{in} \times \frac{R_2}{R_1 + R_2}$$

POTENTIAL DIVIDER

$R = 4\Omega + 5\Omega + 6\Omega = 15\Omega$

RESISTORS IN SERIES

$\frac{1}{R} = \frac{1}{2} + \frac{1}{4} + \frac{1}{12} = \frac{10}{12}$ $R = \frac{12}{10} = 1.2\Omega$

RESISTORS IN PARALLEL

electricity (generators and transformers)

electromagnetic induction is the creation of an **electromotive force** (e.m.f.) in a **conductor** which is moving in a **magnetic field**, or is placed in a changing magnetic field. *Current only flows when the wire is cutting through magnetic field lines. If the wire is stationary, or moves parallel to the magnetic field lines, there is no induced e.m.f.*

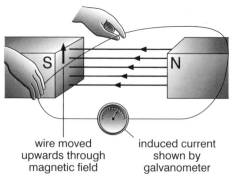

wire moved
upwards through
magnetic field

induced current
shown by
galvanometer

ELECTROMAGNETIC INDUCTION

Faraday's law of induction states that the size of an induced **electromotive force** (e.m.f.) in a conductor is directly proportional to the rate at which the magnetic field changes. *This induced e.m.f. can be increased by:*

● *increasing the speed of the wire's movement.*
● *increasing the length of the wire in the magnetic field.*
● *increasing the strength of the magnetic field.*

generator A generator is a device which produces electrical energy from mechanical energy. *A flat coil of conducting wire is rotated in a magnetic field. This induces an **e.m.f.** in the wire and a current flows.*

dynamo A dynamo is a **generator** which produces electrical energy in the form of **direct current**.

alternator (or **a.c. generator**) An alternator is a **generator** which produces electrical energy in the form of **alternating current**. *The direction of the induced current changes at regular intervals, producing an **alternating current**. Increasing the speed of rotation increases the **frequency** of the alternating current generated.*

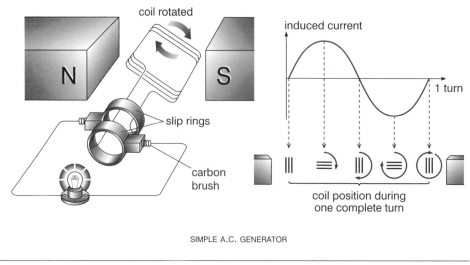

coil rotated

induced current

N

S

slip rings

carbon
brush

1 turn

coil position during
one complete turn

SIMPLE A.C. GENERATOR

generating mains electricity Mains electricity is generated in power stations using large generators, usually powered by heat energy from burning fossil fuels or from nuclear fuels. *This heat boils water to produce steam, which drives a **turbine**, which produces the rotary motion needed for the generator. The electricity produced is transmitted around the country by the **national grid system**.*

national grid system The national grid is a network of overhead cables on pylons and underground cables, for transmitting electricity around the country. *The grid supplies **alternating current** because it uses **transformers** which only work with a.c. The electricity is transmitted at very high voltage (low current) so that less energy is wasted as heat.*

mutual induction is the induction of an e.m.f. in a coil of wire by changing the current in a nearby coil. *The changing current (alternating current) in the first coil produces a changing magnetic field, which induces a current in the second coil. Mutual induction occurs in **transformers**.*

primary coil is the input coil in a **transformer**.

secondary coil is the output coil in a **transformer**.

transformer A transformer is a device for changing the voltage of an **alternating current** without changing its frequency. *It consists of two coils of wire wound on to the same soft-iron **core**. The current in the **primary coil** causes an alternating magnetic field in the iron core. This induces a current in the **secondary coil**. The two coils are not electrically connected and the **power** (voltage times current) in each coil is the same.*

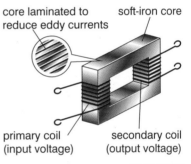

core laminated to reduce eddy currents

soft-iron core

primary coil (input voltage)

secondary coil (output voltage)

TRANSFORMER

eddy currents are currents induced in a piece of metal when the magnetic field around it changes. *Eddy currents are a nuisance as they produce heat and waste energy. In transformers the soft-iron core is laminated (layered into thin varnished sheets) to reduce eddy currents.*

turns ratio The turns ratio is the ratio of the number of turns on the primary and secondary coil of a transformer. *It is equal to the ratio between the voltages of each coil.*

step-up transformer is one in which the number of turns on the secondary coil is greater than the primary coil, so the secondary voltage is greater than the primary voltage. *The **turns ratio** is less than one.*

step-down transformer is one in which the number of turns of the secondary coil is less than the primary coil, so the secondary voltage is less than the primary voltage. *The **turns ratio** is greater than one.*

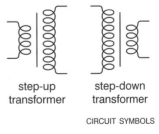

step-up transformer

step-down transformer

CIRCUIT SYMBOLS

direct current (abbreviation: **d.c.**) is an **electric current** which is flowing in one direction only. *All battery-operated electrical devices (torches, calculators, remote controls, etc.) use direct current.*

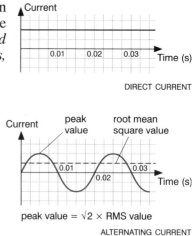

alternating current (abbreviation: **a.c.**) is an **electric current** which reverses its direction of flow in periodic cycles. *In Britain, mains electricity alternates at 50 cycles per second (frequency 50 Hz) and has a voltage of about 230 V.*

peak value = √2 × RMS value

ALTERNATING CURRENT

root mean square value (or **rms value**) is the value of an **alternating current** which would give the same electrical power as a similar d.c. value.

electrical energy is a form of energy which is carried by electric currents, and can be changed into other forms such as heat and light using various electrical appliances. *The amount of electrical energy depends on how many electrons are flowing per second (current) and how much energy each is carrying (voltage). One joule of electrical energy is used when a current of 1 amp flows for 1 second (1 coulomb of charge) under a potential difference of 1 volt. This electrical energy may be converted into heat or light.*

> **energy (J) = potential difference (V) × current (A) × time (s)**

electrical power is the rate at which electrical energy is converted into other forms, e.g.

$$\textbf{power (W)} = \frac{\textbf{electrical energy used (J)}}{\textbf{time (s)}}$$

*Power is measured in **watts** or **kilowatts** (1 kW = 1000 W) or **megawatts** (1MW = 1 000 000 W).*

watt (symbol: **W**) A watt is the SI unit of **power** which is equal to 1 joule per second. *It is a measure of how quickly energy is being transferred and can be linked with brightness (100 W or 60 W bulbs) or amount of heat (1 kW or 2 kW electric fires) With **electrical power** this is equivalent to 1 amp flowing under a potential difference of 1 volt. The unit was named after a British engineer, James Watt (1736–1819).*

kilowatt-hour (symbol: **kWh**) A kilowatt-hour is the amount of **electrical energy** used by a 1 kilowatt device in 1 hour. *It is the commercial unit of electricity and is recorded by electricity meters (1 kW = 3 600 000 J).*

household circuits can be classified into three main types: cooker circuit, ring main, and lighting circuits. *Each of these circuits has a different **fuse rating** (e.g. cooker 30 A, ring main 15 A, lighting 5 A). Appliances are* *arranged in these circuits in **parallel**, so that if an appliance breaks or is switched off, other appliances in the circuit can still work.*

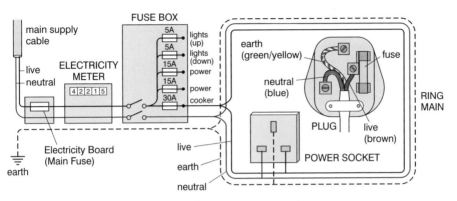

electricity cables (or **power cables**) usually have three insulated wires inside: live (brown wire), neutral (blue wire) and earth (yellow/green wire). *The live and neutral wires carry the current and the earth wire is a safety device. If there is a fault (for example, if the insulation around the live wire is worn) then the metal body of an appliance could become live and give a dangerous shock to anyone touching it. The earth wire 'earths' this current and blows the **fuse**.*

double insulation is produced by enclosing an appliance in a plastic casing so that no metal parts are exposed. *Such appliances do not need an earth wire, and so only require two-pin plugs and sockets.*

switch A switch is a safety device placed on the live wire to switch circuits off when not in use.

fuse A fuse is a short thin piece of wire which overheats and melts to break the circuit if more than a certain value of current flows through. *Fuses, like switches, are always placed on the live wire. All three-pin plugs are fused.*

fuse ratings specify the maximum amount of current that can pass before the fuse overheats and melts (e.g. 3A, 5A, 13A, or 30A). *A suitable fuse must be used for each electrical appliance.*

circuit breakers are a popular alternative to fuses and automatically switch off if a large surge of current passes. *They*

Electrical appliance	Power (watts)	Fuse rating
cooker	8000	30A
immersion heater	3000	13A
kettle	2400	13A
iron	800	5A
colour TV	120	3A
table lamp	60	3A

can easily be 'flicked' back on, which is much easier than replacing the fuse wire in a fuse box, especially if the lights have gone out!

earth leakage is a type of sensitive switch which cuts off the mains current if any current is detected along the earth wire.

electrolysis

electrolysis is the process by which an electric current flowing through a liquid containing **ions** causes the liquid to undergo chemical decomposition. *The electric current is carried not by electrons, but by the movement of the ions.*

electrolytes are liquids which conduct electricity. *All ionic compounds when molten or in aqueous solution are electrolytes, as their ions are free to move. (Liquid metals, in which the conduction is by free electrons, are not regarded as electrolytes.)*

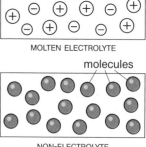

ions

MOLTEN ELECTROLYTE

non-electrolytes are liquids which do not conduct electricity. *Such liquids contain covalent molecules which cannot carry electric current.*

molecules

NON-ELECTROLYTE

weak electrolytes are liquids with a low concentration of ions. *Weak acids and alkalis are weak electrolytes as their aqueous solutions are only partially ionised.*

WEAK ELECTROLYTE

strong electrolytes are liquids with a high concentration of ions. *Strong acids and alkalis are strong electrolytes as their aqueous solutions are fully ionised.*

anions are atoms or molecules containing more electrons than protons, and so carrying a **negative charge**.

STRONG ELECTROLYTE

cations are atoms or molecules containing fewer electrons than protons, and so carrying a **positive charge**.

electrode An electrode is a piece of metal or carbon (graphite) placed in an electrolyte which allows electric current to enter and leave during electrolysis.

anode The anode is a positive **electrode** to which the **anions** (negative ions) are attracted during electrolysis.

cathode The cathode is a negative **electrode** to which the **cations** (positive ions) are attracted during electrolysis.

ionic theory of electrolysis The ionic theory explains how electricity passes through an electrolyte. *During electrolysis, cations travel to the cathode where they gain electrons to form atoms. Anions travel to the anode where they lose electrons to form atoms. This results in an overall movement of electrons from the cathode to the anode, which constitutes an electric current.*

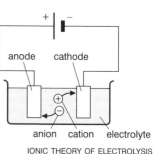

anode   cathode

anion   cation   electrolyte

IONIC THEORY OF ELECTROLYSIS

preferential discharge (or **selective discharge**) occurs because ions which require the least amount of energy are discharged first. *Cations (metal ions) higher in the reactivity series are harder to discharge at the cathode than those lower down. There is also an order for ease of discharge for anions (see table).*

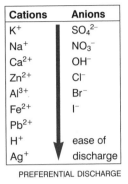

Cations	Anions
K^+	SO_4^{2-}
Na^+	NO_3^-
Ca^{2+}	OH^-
Zn^{2+}	Cl^-
Al^{3+}	Br^-
Fe^{2+}	I^-
Pb^{2+}	
H^+	ease of
Ag^+	discharge

PREFERENTIAL DISCHARGE

Hoffmann voltameter A Hoffmann voltameter is a type of electrolytic cell which is used to collect and measure volumes of gases liberated during electrolysis. *Electrolysis of acidified water produces twice as much hydrogen as oxygen. This indicates a ratio of 2:1 in the chemical composition of water* (H_2O).

Faraday's laws of electrolysis are two laws named after the British scientist Michael Faraday (1791–1867).

First law: The amount of chemical change during electrolysis is directly proportional to the quantity of electrical charge passed.

Second law: The amount of chemical change during electrolysis is inversely proportional to the charge on an ion.

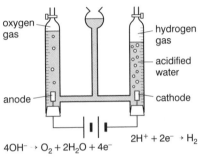

oxygen gas — hydrogen gas — acidified water — anode — cathode

$4OH^- \rightarrow O_2 + 2H_2O + 4e^-$

$2H^+ + 2e^- \rightarrow H_2$

HOFFMANN VOLTAMETER

electroplating is the coating of a metal object with a thin layer of another metal by electrolysis. *The object to be plated is made the **cathode**, and the plating metal is made the **anode**. The electrolyte is an aqueous solution of a salt of the plating metal.*

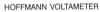

electrolyte: solution of metal salt

anode: plating metal (e.g. Zn, Ni, Cr, Cu) — cathode: object to be plated

ELECTROPLATING

electrorefining (or **electrolytic refining**) is a method of purifying metals such as copper by electrolysis. *The impure copper is made the **anode**. Copper ions dissolve and travel from the anode to the cathode where pure copper builds up. Impurities do not dissolve, and fall to the bottom as 'anode sludge'. This sludge may contain valuable impurities (silver, gold) which are extracted by further electrolysis.*

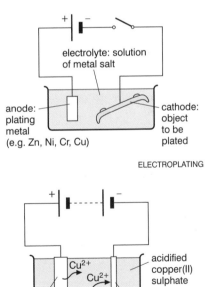

acidified copper(II) sulphate electrolyte

impure copper anode
$Cu \rightarrow Cu^{2+} + 2e^-$

anode sludge

pure copper cathode
$Cu^{2+} + 2e^- \rightarrow Cu$

COPPER REFINING

aluminium extraction: see **metals (extraction)**

electromagnetic waves

electromagnetic waves are **transverse waves** produced by oscillating electric and magnetic fields at right angles to one another. *They do not require a medium in which to propagate and can travel through a **vacuum**. As waves, they undergo **reflection, refraction**, and **diffraction**. For example,* radio waves of long and medium wavelengths can be reflected off the **ionosphere**. *This allows signals to travel great distances without being blocked by the curvature of the Earth. They can also be bent by diffraction, e.g. when travelling over a hill.*

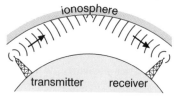

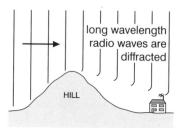

REFLECTION OF RADIO WAVES

speed of light (symbol: c) All electromagnetic waves travel at the same speed in a vacuum, which is approximately $3 \times 10^8 \, \mathrm{m\,s^{-1}}$ or $300\,000 \, \mathrm{km\,s^{-1}}$.

frequency The frequency of a wave is the number of oscillations per second. *It is related to wavelength by the equation $c = f \times \lambda$, where c is the velocity of light, f is the frequency of the wave and λ is the wavelength. As c is a constant, it follows that high-frequency waves have short wavelength and low-frequency waves have long wavelength. High-frequency waves also have greater energy and are therefore more penetrating.*

DIFFRACTION OF RADIO WAVES

inverse square law Waves emitted from a point source in a vacuum obey the inverse square law. *If you double the distance from the source, the intensity of the wave becomes a quarter of its previous value.*

interaction of waves with matter The waves at each end of the spectrum tend to pass through materials. *Waves nearer the middle of the spectrum tend to be absorbed. When the waves are absorbed by **metals** they produce heat, and also create a tiny alternating current with the same frequency as the radiation.*

polarisation occurs when a wave oscillates in one plane only. *Only transverse waves can be polarised.*

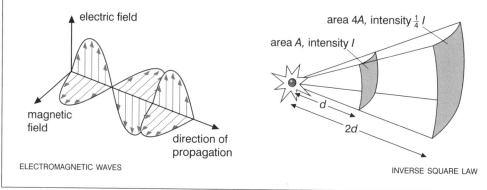

ELECTROMAGNETIC WAVES

INVERSE SQUARE LAW

electromagnetic spectrum The electromagnetic spectrum is the range of frequencies over which electromagnetic waves are propagated. *Although the spectrum is continuous it can be split into seven overlapping regions (see table).*

	gamma rays	X-rays	ultraviolet (UV) radiation	visible light	infra-red (IR) radiation	microwaves	radio waves
Wavelength (m)	10^{-13}–10^{-12} 10^{-11} 10^{-10}	10^{-9}	10^{-8} 10^{-7}		10^{-6} 10^{-5} 10^{-4} 10^{-3} 10^{-2} 10^{-1}		1 10^{1} 10^{2} 10^{3} ···
Frequency (Hz)	10^{21} 10^{20} 10^{19}	10^{18}	10^{17} 10^{16} 10^{15}		10^{14} 10^{13} 10^{12} 10^{11} 10^{10} 10^{9}		10^{8} 10^{7} 10^{6} 10^{5} ···
Sources	radioactive substances (uranium)	X-ray tubes	very hot objects (the Sun, mercury vapour lamps)	hot objects (the Sun)	warm or hot objects (fires, living bodies)	microwave ovens	radio transmitters (including radar and television transmitters)
Detection	Geiger–Müller tubes	special photographic film	fluorescent material which absorbs rays and changes them into visible light	eyes	easily absorbed by most objects causing a rise in temperature	cause molecules to vibrate and become very hot	radio aerials, TV aerials
Uses	• cancer treatment • sterilisation of equipment	• (at low energy) images of internal organs	• washing powders use fluorescent chemicals • sun-tan lamps • invisible marking for security	• optic fibres • seeing • photography	• thermal images images of body in medicine • infra-red cameras for seeing objects at night • remote controls for TVs, VCRs	• microwave ovens • communication (satellite TV, mobile phones)	• communication systems (radio, terrestrial TV)

increasing wavelength →

increasing frequency →

electromagnetism

electromagnetism is the combination of an **electric field** and a **magnetic field** and their interaction to produce a force. *When an electric current flows in an insulated straight wire, it produces a circular magnetic field around the wire. This field can be plotted using a small compass. The direction of the field depends upon the direction of the current in the wire. Electromagnetism is important in **electric motors**, in **generators**, and in making powerful **electromagnets**.*

coil A coil is a number of turns of insulated wire carrying a current. *It is made by wrapping the wire around a shaped piece of material called the former. Both flat coils and **solenoids** are used in **generators** and **motors**.*

solenoid A solenoid is a long cylindrical **coil** of insulated wire. *A current flowing through a solenoid produces a magnetic field which is similar to that produced by a bar magnet. The position of the poles depends on the direction of the current. The magnetic field of a solenoid can be increased by increasing the current and/or the number of turns of conducting wire.*

core The core is the material in the centre of a coil or solenoid which increases the strength of the field. *It is normally a rod made of a **ferromagnetic material** such as soft iron (pure iron).*

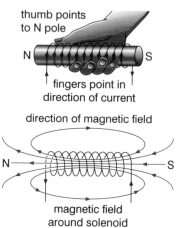

thumb points to N pole

N   S

fingers point in direction of current

direction of magnetic field

N   S

magnetic field around solenoid

SOLENOID

electromagnet An electromagnet is a **solenoid** with a **core** of ferromagnetic material such as soft iron. *This forms a temporary magnet which can be switched on and off simply by switching the current on and off. Electromagnets have many uses as switches and in turning electrical energy into mechanical energy.*

relay (or **electrical relay**) A relay is a device which uses a small current in the coil of an **electromagnet** to switch on a large current in another independent circuit. *Relays are safer than ordinary switches and have a wide range of uses in electrical and electronic circuits.*

pivot

soft-iron core

+ •

contacts

–

+

main circuit

(large current)

electromagnet

small current

RELAY

electric bell An electric bell is a device which uses an **electromagnet** to operate a hammer striking a bell. *It uses direct current which, when the switch closes, activates the electromagnet, which then pulls the hammer on to the bell. This breaks the circuit, which switches the electromagnet off, so that the hammer returns. The process repeats itself, with the hammer repeatedly striking the bell, until the bell push is released.*

the motor effect is that when a wire carrying a **current** is brought into a **magnetic field** there is repulsion between the magnetic field of the current and the field of the magnet, which causes a **force** on the wire. *The size of this force can be increased by increasing the current and/or the strength of the magnetic field. The direction of this force is indicated by Fleming's left-hand rule.*

Fleming's left-hand rule (or **motor rule**) gives the direction of the motor effect. *The thumb and first two fingers of the left hand are held at right angles to each other. The thumb shows the direction of motion. The first finger points to the direction of the magnetic field and the second finger shows the direction of the current.*

electric motor An electric motor is a device which uses the **motor effect** to change electrical energy into mechanical energy. *A simple d.c. motor consists of a flat coil of current-carrying wire placed in a magnetic field. One side of the coil experiences an upward force, the other side a downward force, so the coil rotates to produce mechanical motion.*

commutator A commutator is a device used in a d.c. **electric motor** to reverse the current direction every half turn. *It is made of a split metal ring so that the current will enter and leave the commutator through two carbon brushes.*

loudspeaker A loudspeaker is a device which uses the **motor effect** to change electrical energy into sound energy. *It consists of a **coil** of wire in a radial **magnetic field**. The coil is attached to a paper cone. The changing current in the coil makes the paper cone vibrate, producing sound.*

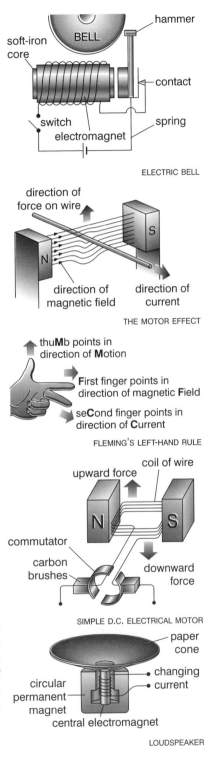

ELECTRIC BELL

THE MOTOR EFFECT

FLEMING'S LEFT-HAND RULE

SIMPLE D.C. ELECTRICAL MOTOR

LOUDSPEAKER

electronics is a branch of physics and technology concerned with the study and use of small electric currents passed through semiconductors and gases at low pressure. *Many electronic components are used as switches in circuits because their ability to conduct electricity can be affected by factors like current direction (see **diode**), temperature (see **thermistor**), and light (see **light-dependent resistor**).*

diode A diode is an electronic device made from a **semiconductor** material such as silicon which can be used as a one-way switch. *Diodes have very low resistance in one direction and very high resistance in the other. They can be used to change **alternating current** to **direct current**.*

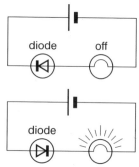

transistor A transistor is an electronic device which is commonly used as a switch or an amplifier. *Transistors are connected into electronic circuits at three points called the base, collector, and emitter. When a small current (about 5mA) flows to the base, the resistance between the collector and emitter changes from very high to very low and the transistor is switched on. By varying the size of this current to the base we can control when the transistor switches on, and therefore when the **electronic switch** is activated.*

light-emitting diode (abbreviation: **LED**) A light-emitting diode is a semiconducting diode that has a higher resistance than normal and produces light instead of heat. *LEDs, like many electronic components, work with very small currents. They are widely used for displaying letters and numbers in digital instruments in which a self-luminous display is needed, e.g. calculators, watches, and hi-fi equipment.*

LED

light-dependent resistor (abbreviation: **LDR**) A light-dependent resistor is a resistor made from a **semiconductor** (e.g. cadmium sulphide or selenium) whose resistance changes with light intensity. *In the dark LDRs have very high resistance, but in the light a low resistance. They can be used as light-dependent switches for opening automatic doors or in alarm systems.*

LDR

thermistor A thermistor is a resistor made from a **semiconductor** whose resistance falls sharply when its temperature rises above room temperature. *Such resistors can be used as temperature switches which switch on devices when they have 'warmed up'.*

THERMISTOR

electronic switches are **transistors** and other electronic components which can be activated under different conditions. *A **potential divider** is used to vary the voltage across the base and the emitter of the transistor.*

Resistors R_1 and R_2 (see diagram) form the potential divider as they divide the voltage from the battery. When $R_2 \gg R_1$ there will be a high voltage between the base and the emitter and a low voltage and small current going to the base. The transistor will switch on.

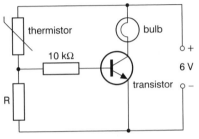

POTENTIAL DIVIDER ACTS AS AN ELECTRONIC SWITCH

heat-sensitive switch A heat-sensitive switch is an electronic switch which responds to changes in temperature. *The **thermistor** has a high resistance at low temperature and therefore has a high voltage across it. This means that there is a low voltage across the bottom resistor R, so there is no current from the base to the emitter and the transistor and bulb are switched off. When the temperature rises, the thermistor's resistance falls and the situation is*

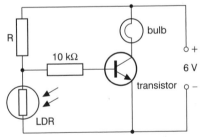

HEAT-SENSITIVE SWITCH

reversed. The potential difference across resistor R becomes large enough for current to flow from the base to the emitter, so the transistor switches on and the warning bulb lights up.

light-sensitive switch A light-sensitive switch is an electronic switch which responds to changes in light intensity. *The **light-dependent resistor (LDR)** has a high resistance in the dark and therefore has a high voltage across it. In the dark, the potential difference becomes large enough for current to flow from the base to*

LIGHT-SENSITIVE SWITCH

the emitter, and the transistor switches on, so the bulb lights up. In the light, the transistor and lamp are switched off as the potential difference between the base and the emitter is very low.

electronic systems typically consist of input sensors which feed information into a processor, which is then connected to an output device controlled by the processor.

silicon chips are thin wafers of pure silicon with miniature electronic circuits printed on them. *These printed integrated circuits range from simple logic circuits (logic gates) to chips about 8 mm square which contain a million or so components.*

energy is the capacity of a system to do **work**. *The word is used in everyday life to describe the ability to do something useful. When work is done on or by an object then the object gains or loses energy respectively. Energy exists in various forms and is only useful when converted from one form to another. All forms of energy are measured in* **joules**.

work is the energy transfer that occurs when a force causes an object to move a certain distance in the direction of the force.

> work (J) = force (N) × distance moved in the direction of the force (m)

Work can only be done if a force moves something. Energy can be used up without work being done if the object does not move. For example, work is done when a crate is lifted. However, if the crate is too heavy to lift, then no work is done, but energy is still used up in trying to lift the crate. Work is measured in **joules**.

joule (symbol: **J**) A joule of work is done by a force of one **newton** moving one **metre** in the direction of the force. *The joule is the SI unit of both* **work**, *and* **energy**. *The unit is named after the British physicist James Joule (1818–89).*

forms of energy There are many different manifestations of energy. Some important forms are **electricity**, **heat energy**, **sound**, **electromagnetic waves** and **nuclear energy**. *(These have their own sections.)*

potential energy is energy which is stored in a body or system because of its position, shape, or state.

gravitational potential energy (abbreviation: **GPE**) is the stored energy an object has because of its position above the Earth. *The further away from the Earth, the greater the potential energy. The relationship can be expressed by the following equation:*

> potential energy (J) = mass (kg) × gravitational acceleration × height (m)

An object of 1 kg mass which is 1 metre away from the Earth's surface has a potential energy of approximately 10 J (assuming g = 10 ms^{-2}). When an object falls to the Earth it loses its potential energy, which is changed into kinetic energy of motion.

elastic potential energy (or **strain energy**) is the stored energy an object has as a result of stretching or compressing. *Only certain types of object can store elastic potential energy, such as stretched elastic bands and squashed springs.*

chemical potential energy is the energy stored in systems such as fuel and oxygen, food and oxygen, and chemicals in **batteries**. *This* **chemical energy** *is released during chemical reactions such as the burning of fuels or the respiration process.*

kinetic energy (or **moving energy**) is the energy possessed by an object or particle because it is moving. *The greater the mass of the object and/or the greater its velocity, the greater its kinetic energy. The relationship can be expressed by the equation:*

$$\text{kinetic energy (J)} = \frac{1}{2} \times \text{mass (kg)} \times \text{(velocity)}^2$$

mechanical energy is the sum of the **kinetic energy** and the **gravitational potential energy** of an object.

energy conversion (or **energy transfer**) is a change of one energy form into another. *The original source of energy is often the Sun. The Sun's energy can be stored as **chemical potential energy** in fossil fuels and food. This can be released by burning or respiration to provide kinetic and heat energy, which can then be converted to electrical energy, sound energy, etc. Electrical energy is useful because it is relatively easy to convert to other forms of energy.*

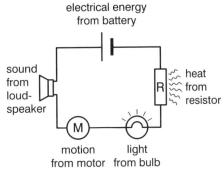

CONVERSION OF ELECTRICAL ENERGY

transducer A transducer is a device which converts electricity into different energy forms (sound, light, mechanical, etc.) or vice versa.

law of conservation of energy This law states that energy cannot be created or destroyed, but can be converted from one form to another. *In any **energy conversion** the total number of joules of energy at the beginning and the end is the same but the quality of energy becomes less because some of it is converted to waste heat (random thermal energy).*

Transducer	Main energy conversion
battery	chemical → electrical
light bulb	electrical → light
stereo	electrical → sound
thermocouple	heat → electrical
solar cell	light → electrical
speaker	electrical → sound
microphone	sound → electrical
dynamo	mechanical → electrical
motor	electrical → mechanical

power is the rate at which work is done or energy is transferred. *The power of a person or a machine is the rate at which they change one form of energy into another. Power is measured in **watts** (joules per second) and can be calculated by the equation:*

$$\text{power (W)} = \frac{\text{work done (J)}}{\text{time taken (s)}}$$

*Both mechanical power and electrical power use the same unit (see **electrical power**).*

enzymes and biotechnology

enzyme An enzyme is a **protein** which alters the rate of a particular biochemical reaction. *Enzymes are often called 'biological catalysts'. Enzymes are produced in living **cells** and work mostly in the **cytoplasm**. They work best at the normal temperature of the living cell, and within a narrow range of pH, which may be different for each enzyme.*

denaturing Above 50°C the molecular shape of an enzyme is changed, and it becomes denatured and stops working.

substrate A substrate is a molecule on which an enzyme acts in a biochemical reaction. *The enzyme **amylase** catalyses the breakdown of **starch** into sugar by the reaction of water (hydrolysis). The substrate in this reaction is the starch molecule, which forms an enzyme–substrate complex with the amylase during the reaction.*

active site The active site is an area on the **enzyme** molecule to which the **substrate** attaches during the reaction. *The **active site** is very specific and acts like a lock for a key. Only the specific substrate will fit the active site.*

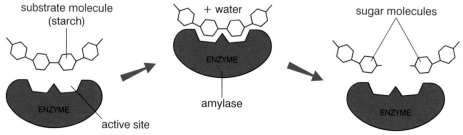

BREAKDOWN OF STARCH BY THE ENZYME AMYLASE

immobilised enzymes are insoluble enzymes which can be re-used in industrial processes. *An enzyme which does not dissolve but still allows the substrate to react with it can then be used again, which reduces waste. Such enzymes are called 'immobilised enzymes' as they are fixed in insoluble substances on permanent supports. The substrate flows over them and the enzyme catalyses the reaction.*

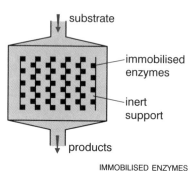

IMMOBILISED ENZYMES

biotechnology is the use of living organisms (containing enzymes) for the production of useful substances or processes. *Examples are **brewing**, **baking**, **cheese making**, **sewage** treatment, and **genetic engineering**, which can be used to modify bacterial cells to produce substances such as **hormones** (e.g. insulin), **vaccines**, etc.*

yeasts are a group of unicellular **fungi** which carry out the biochemical process used in **baking** and **brewing**. *They are **saprophytes** which secrete enzymes to convert sugars into other substances.*

brewing is the use of brewer's **yeast** to convert sugar solution into **alcohol** by a **fermentation** reaction. *The main sugar is maltose from germinating barley seeds.*

baking is the use of baker's **yeast** mixed with flour and water (dough) to make bread. *Flour contains **starch** which, when mixed with water, is digested by the yeast and produces bubbles of carbon dioxide gas. These get trapped inside the dough and make it rise. Flour also contains a protein called gluten. This forms sticky threads as the bread is kneaded and helps to trap the carbon dioxide. When the bread is baked the yeast is killed, the alcohol evaporates, and the carbon dioxide escapes. The remaining starch and gluten give the bread its firm texture.*

cheese making is the conversion of milk into cheese by a controlled process using natural bacteria. *Milk naturally contains **bacteria** (lactobacillus) which perform **anaerobic respiration** when provided with a source of sugar. Milk also contains the sugar lactose, which is converted by these bacteria into lactic acid and energy. The presence of the acid makes the milk turn sour, as the proteins in the milk coagulate and form clumps called curds in a watery liquid called whey. An enzyme called rennin may be added, to speed up this process. The curds are then separated and pressed to make cheese.*

sewage is waste water from homes and factories which contains excretory waste (faeces, urine), used washing water, and surface water. *Raw (untreated) sewage is a pollutant and must be treated before being discharged. Sewage treatment involves **filtration**, **sedimentation**, **digestion**, and **aeration**.*

filtration of sewage involves passing it through screens to remove floating debris and waste.

sedimentation of sewage is carried out by leaving it in tanks so that tiny insoluble particles collect at the bottom as sludge.

digestion in sewage treatment uses the action of bacteria in a 'sludge digester' to break down organic matter, producing methane gas and solid material ('sludge cake') which can be dried as **fertiliser**.

aeration is the process by which clear water from the top of **sedimentation** tanks is sprayed over filter beds, where it dissolves oxygen from the air. *Harmful anaerobic bacteria are killed, and useful aerobic bacteria multiply. The treated water is then returned to the nearest river or the sea.*

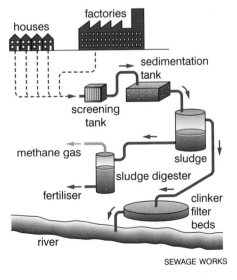

SEWAGE WORKS

evolution is the gradual changing of a species of living organism over a long period of time.

Darwinism is a theory of evolution which states that present-day living creatures have developed by gradual changes over many generations, as a result of *natural selection. This theory of evolution was proposed by the British naturalist Charles Darwin (1809–82), in his book* On the Origin of Species, *published in 1859.*

natural selection The theory of natural selection states that the individual organism which is best adapted to its environment will survive to reproduce. *This is the basic theory of **Darwinism**. It can be summarised as follows:*

- **overpopulation** Most organisms produce more young than will survive to adulthood.
- **variation** Within any population of organisms there are slight variations. Some variations may better adapt the organism for survival.
- **survival of the fittest** Overpopulation causes competition in which only the fittest will survive.
- **inheritance** Organisms which have an advantageous characteristic are more likely to survive and reproduce. This advantageous variation will then be passed to its offspring, which will also stand a better chance of survival.
- **adaptation** Gradually over a period of time, each generation of a particular organism will become better adapted to its environment.

evidence for evolution comes from **fossils**, and from **homologous** and **vestigial structures**. *Comparison of fossil records show the gradual change over millions of years. Fossils can be accurately dated by **radioactive carbon dating**. However, there are many missing links, as most organisms decay almost completely and leave no fossil remains. Very occasionally an organism is found completely preserved (e.g. a mammoth in a glacier, or an insect in the fossilised resin called amber).*

homologous structures are structures in different organisms which have fundamental similarities although they may have developed quite different functions. *For example, the forelimbs of all vertebrates have similar bone structure.*

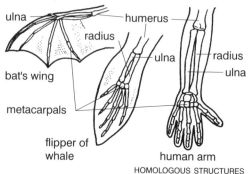

ulna — humerus — radius — bat's wing — ulna — radius — ulna — metacarpals — flipper of whale — human arm

HOMOLOGOUS STRUCTURES

vestigial structures are reduced structures in a plant or animal which serve no function but have been 'left over' after the evolutionary process. *Examples are the **appendix** in humans, the wings of flightless birds such as the ostrich, and the limb girdles of snakes.*

variation is the range of differences between members of a species. *Only differences which are genetic in origin can be inherited and so acted on by natural selection.*

continuous variation shows a range of values between two extremes. *Height, weight, hair colour, skin colour, IQ all show continuous variation. A graph showing such a variation is usually a normal distribution curve. Most of the population come in the middle of the range with few at the lower and upper ends. Continuous variation is usually controlled by many different genes.*

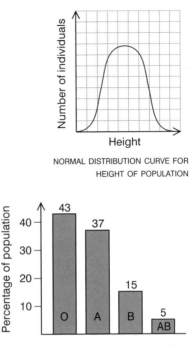

NORMAL DISTRIBUTION CURVE FOR HEIGHT OF POPULATION

discontinuous variation consists of specific characteristics or values with nothing in between. *Blood groups show discontinuous variation. You are either A, B, O, or AB. Other examples: garden peas are either wrinkled or smooth; people have earlobes either attached or unattached. Discontinuous variations are often controlled by one particular gene.*

BLOOD GROUP IS A DISCONTINUOUS VARIABLE

mutation A mutation is a sudden random change in the genetic material of a cell, which may result from faulty DNA replication or faulty division of chromosomes. *If a mutation occurs in a sex cell, then it may be passed on to the next generation. Most mutations are harmful, but some increase the 'fitness' of the organism to survive, e.g. when bacteria become resistant to antibiotics. Mutation is essential to **evolution** because it is the ultimate source of **genetic variation**. The likelihood of mutation is increased by **radiation** (ultraviolet rays, X-rays, and gamma rays) and by chemicals called mutagens, e.g. mustard gas, nitrosamines, and other carcinogens (cancer-causing chemicals).*

acquired characteristics are physical characteristics which are acquired by an individual organism during its lifetime. *Such characteristics are not genetic and are not passed on to future generations, e.g. scars following wounds or the stunted growth of a plant growing in poor soil.*

Lamarckism is an early theory of evolution proposed by the French biologist Jean-Baptiste de Lamarck (1744–1829). *The theory has been rejected and superseded by **Darwinism**. Lamarck thought that **acquired characteristics** during an individual's lifetime are passed on to its offspring. He explained the long neck and limbs of a giraffe as having evolved by the animal stretching its neck to reach the foliage of trees. It is now accepted that **variation** in the giraffe resulted in the **natural selection** of long necks and limbs which, when passed on to the offspring, improved its chances of survival.*

excretion and the urinary system

excretion is the removal of waste products formed as a result of biochemical reactions inside a living organism. *Excretion is important in maintaining a constant internal environment in the organism (**homeostasis**). In mammals, the important excretory organs are the **lungs** (excrete carbon dioxide), **skin** (excretes sweat), and **kidneys**.*

urinary system The urinary system is the main system in the body concerned with the removal of waste material from the blood (**excretion**) and with water regulation (**osmoregulation**). *In mammals the urinary system consists of two **kidneys** each linked to the **bladder** by a **ureter**.*

osmoregulation is the control of water content and the concentration of salts in an animal's body.

kidney The kidney is the main excretory organ in vertebrates. *In humans a pair of kidneys are situated at the back of the body, just below the ribs. They remove unwanted substances from the blood and regulate the amount of water and salts in the body.*

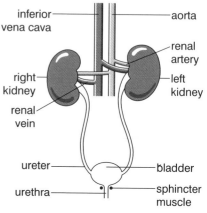

HUMAN URINARY SYSTEM

renal artery is the artery that takes blood to the kidneys.

renal vein is the vein that takes blood away from the kidneys.

urine is a liquid produced in the kidneys and stored in the bladder. *It is an aqueous liquid containing nitrogenous waste materials such as **urea**, uric acid, and ammonia.*

urea is a toxic compound produced when proteins are broken down.

ureters are two tubes which carry urine from the kidneys to the bladder.

urethra The urethra is the tube through which urine is discharged from the bladder to the exterior (urination).

bladder (or **urinary bladder**) The bladder is a hollow muscular organ which stores urine before it is discharged. *Flow of urine from the bladder is controlled by a ring of muscle (urinary sphincter) between the bladder and the urethra.*

structure of the kidney A longitudinal section through the kidney shows three main parts:

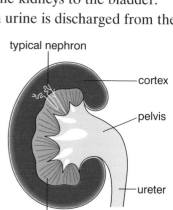

LONGITUDINAL SECTION OF KIDNEY

- **cortex** (or **renal cortex**) is the outermost layer of tissue in the kidney.
- **medulla** (or **renal medulla**) is the central tissue of the kidney. The outer cortex and

the inner medulla are made up of thousands of tiny tubules called nephrons.
- **pelvis** (or **renal pelvis**) is the cavity in the kidney that receives urine from the nephrons which then drains into the ureter.

nephron (or **kidney tubule**) A nephron is the filtering unit in the kidney. *Through these nephrons, nitrogenous waste is filtered from the blood with the formation of* **urine**. *This involves the processes of* **filtration** *and* **reabsorption**.

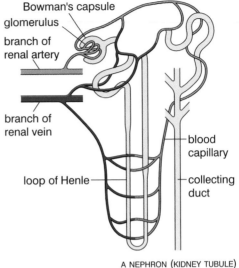

A NEPHRON (KIDNEY TUBULE)

Bowman's capsule A Bowman's capsule is the closed end of a **nephron**, and is found in the **cortex** of the kidney where **filtration** takes place. *Each capsule is shaped like a cup and contains a* **glomerulus**.

glomerulus A glomerulus a tangle of blood capillaries located in each **Bowman's capsule** of the kidney.

filtration occurs in the **Bowman's capsules**. *The renal artery brings blood to each glomerulus, where the pressure builds up as the capillaries are narrow. This pressure pushes small molecules out of the blood capillaries and into the Bowman's capsule. Only small molecules can pass out of the capillaries (water, glucose, amino acids, urea, minerals and vitamins). These make up the glomerular filtrate which passes through the rest of the kidney tubule. The larger molecules like proteins and blood cells remain in the capillaries of the nephron. This filtration under pressure is also called 'ultrafiltration'.*

reabsorption (or **selective reabsorption**) occurs along the length of the **nephron** or tubule. *Wrapped around each nephron are blood capillaries. These reabsorb useful substances (most of the water, glucose, amino acids, vitamins, etc.) from the glomerular filtrate back into the blood. The remaining filtrate, including surplus water, all the nitrogenous waste (urea), and some mineral salts, collects at the end of the nephron and passes as urine to the ureter.*

dialysis is the process of clearing waste substances such as urea and ammonia from the blood of people who have suffered kidney failure, using a 'kidney machine'.

kidney transplants are an alternative to dialysis and are one of the most successful of transplant operations. *Best results occur if the donor has similar tissue and body chemistry (identical twins have very successful transplants). If the donor is unrelated, then the body may reject the new kidney as if it were a disease organism. This can be treated with drugs.*

eye The eye is the organ of sight, which focuses and detects light and passes nerve impulses to the brain via the optic nerve.

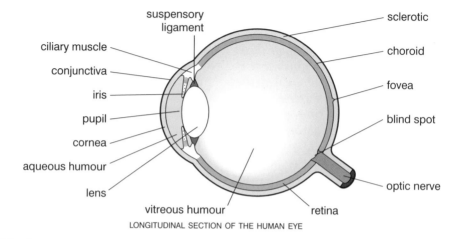

LONGITUDINAL SECTION OF THE HUMAN EYE

orbit (or **eye socket**) The orbit is a deep cavity in the skull in which the eyeball is situated for protection.

extrinsic muscles are the muscles attached to the outside of the eyeball which allow the eye to move and rotate so that it can follow and focus on objects.

conjunctiva The conjunctiva is a thin, transparent, self-repairing membrane at the front of the eye. *It is kept moist by the fluid of tears secreted by the tear gland (or lacrimal gland). This fluid contains an enzyme (lysozyme) which helps to kill bacteria.*

sclerotic (or **sclera**) The sclerotic is the tough white outermost layer of the eyeball. *It encloses the **choroid** and **retina** and is continuous with the **cornea**.*

aqueous humour is a watery liquid secreted and absorbed into the front cavity of the eye. *It is renewed about every four hours. The **cornea** and **lens** obtain food and oxygen by diffusion through the aqueous humour.*

vitreous humour is a jelly-like material which fills the rear cavity of the eye. *This material exerts outward pressure to maintain the shape of the eyeball, and helps to refract (bend) the light.*

retina The retina is the layer of light-sensitive (photoreceptive) cells at the back of the eye. *There are two types of receptor cell called **rods** and **cones**. These send information along the optic nerve to the brain. The brain then interprets the information from each receptor to build up an image.*

rods are light-sensitive cells in the **retina** which are sensitive to quite dim light but not to colours.

cones are light-sensitive cells in the **retina** which are sensitive to bright light and give colour vision.

fovea The fovea is the most sensitive part of the **retina** where the receptor cells (all cones) are packed most closely together.

blind spot The blind spot is an area where the optic nerve and the retina meet. *There are no receptor cells at this point.*

choroid The choroid is the black layer behind the retina which absorbs all the light after it has passed through the retina.

cornea The cornea is the transparent layer at the front of the eye. *It is a continuation of the **sclerotic**, and acts like a lens, bending the light as it passes through. It has no direct blood supply, as a network of capillaries would interfere with the focusing of the light.*

pupil The pupil is the hole at the centre of the **iris** which appears as a black circle. *It allows light to enter and pass through the **lens** to the **retina**. Its size is controlled by the muscles of the **iris**.*

iris The iris is the coloured part of the eye and controls the amount of light that reaches the retina. *It has two types of muscle: circular muscles running around the pupil, which contracts in bright light, and radial muscles running outwards from the edge of the pupil, which contract in dim light.*

lens The lens bends the light as it passes through the eyeball so that the light is focused on the **retina**. *It consists of layers of transparent material. The lens is held in place by suspensory ligaments attached to its outer rim. These are attached to a ring of muscle fibres around the lens called the ciliary muscles.*

accommodation is changing the thickness of the lens to focus on objects at various distances from the eye. *When the ring of ciliary muscles contracts, the suspensory ligaments are loosened and the lens becomes fatter, to focus on nearby objects. When the ciliary muscles are relaxed, the suspensory ligaments are tightened which makes the lens thinner, to focus on distant objects.*

long-sightedness (or **hypermetropia**) is a vision defect which makes people able to focus clearly only on distant objects. *The solution is to wear glasses or contact lenses with **convex** (converging) **lenses**. These bend the rays inwards so that they focus on the retina and not behind it.*

short-sightedness (or **myopia**) is a vision defect which makes people able to focus clearly only on nearby objects. *The solution is to wear glasses or contact lenses with **concave** (diverging) **lenses**. These bend the rays outwards so that they focus on the retina and not in front of it.*

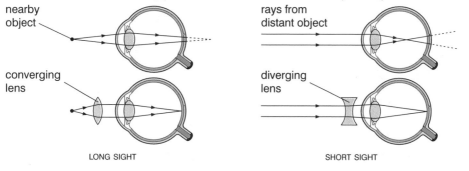

nearby object

converging lens

LONG SIGHT

rays from distant object

diverging lens

SHORT SIGHT

nutrition (or **feeding**) is the process by which living organisms obtain the substances they need to provide materials and fuel for growth, repair of tissues, etc. (It should not be confused with **respiration**, which is the chemical release of energy from food and oxygen.)

autotrophic nutrition (or **holophytic nutrition**) is the making of an organism's own food from inorganic substances, e.g. by plants using **photosynthesis**. *Autotrophs such as plants do not rely on other living organisms for food.*

heterotrophic nutrition (or **holozoic nutrition**) is the feeding of organisms on organic substances made by other organisms. *All animals and fungi, and some bacteria, are heterotrophs.*

nutrients are substances which are essential for healthy growth. *Plants need carbon dioxide (from the air) and water (from the soil) for **photosynthesis** to make **carbohydrates**, and use **minerals** dissolved in the water to make other molecules such as **proteins**. Animals (and humans) need **carbohydrates**, **proteins**, and fats, and also **vitamins** and **minerals** (a **balanced diet**).*

carbohydrates are a group of organic compounds with a general formula of $C_x(H_2O)_y$ and are important nutrients, especially for energy. *Carbohydrates include all **sugars**, **starch**, **glycogen**, and **cellulose**. Foods rich in carbohydrates include sugary foods such as jam, honey, sweets, and cakes, and starchy foods such as potatoes, rice, bread, and spaghetti. Vegetables and cereal foods contain cellulose.*

fats and oils are a group of organic compounds made up of carbon, hydrogen, and a small amount of oxygen, and are an important fuel source, having twice the energy value of carbohydrates. *Fats are solids and oils are liquids at room temperature. Foods rich in fats include butter, lard, margarine, olive oil, etc.*

Food	Energy per 100 g
lard	3700 kJ
sugar	1600 kJ
bread	100 kJ
fish fingers	750 kJ
lettuce	35 kJ

adipose tissue is a special tissue in which the body stores fats. *This tissue occurs under the skin and around the muscles, heart and kidneys.*

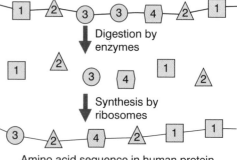

Amino acid sequence in food protein

Digestion by enzymes

Synthesis by ribosomes

Amino acid sequence in human protein

proteins are large organic molecules which are essential for the growth and repair of body tissue. *Protein can also be used as an energy source. Foods rich in protein include meat, fish, eggs, dairy products, beans, whole grains and nuts. Animals take in proteins and break them down by **digestion** into **amino acids**. These are then transported in the blood to body cells where they are* *reassembled by the cells' **ribosomes** to make the different **proteins** the body needs.*

minerals are natural inorganic substances which are needed for building certain body tissues. *Plants depend on minerals in the soil, dissolved in the water and absorbed by the roots. Animals obtain minerals from plant or animal foods.*

Mineral	Needed for	Food source
Calcium	teeth, bones	cheese, milk, vegetables
Iron	making blood	red meat, eggs, bread
Sodium	muscle movement	salt

vitamins are organic compounds required by animals in small amounts to maintain health (see table). *There are two major groups, those that are water soluble (e.g. vitamins B and C) and those that are fat soluble (e.g. vitamins A, D, E, and K). Many vitamins are destroyed by cooking.*

Vitamin	Deficiency disease	Food sources
A (retinol)	weakens vision (night blindness)	carrots, milk
B$_1$ (thiamine)	beri-beri	yeast, beans
B$_2$ (riboflavin)	mouth sores	yeast, liver
C (ascorbic acid)	scurvy	citrus fruit
D (calciferol)	rickets (soft bones)	cod liver oil, eggs
E (tocopherol)	infertility	cereal, green vegetables
K	poor blood clotting	egg yolk, green vegetables

roughage (or **fibre**) is the part of food that cannot be digested. *It is necessary for the proper working of the **alimentary canal**, and helps prevent, appendicitis, constipation, obesity, and cancer of the bowels. Foods rich in fibre include wholemeal cereals, nuts, fruit, and root vegetables.*

balanced diet A balanced diet consists of food which contains carbohydrates, fats, proteins, minerals, and vitamins in the correct proportions to maintain good health. *Balanced diets vary depending on the age, sex, body size, and level of activity of the individual. Poor diet can lead to **malnutrition**. Vegetarians have to be careful that they get enough of certain nutrients which other people get from animal food products.*

obesity is the condition of being very overweight, often because of eating too much carbohydrate and fat. *This can lead to **heart disease, high blood pressure**, etc. Many animal fats can cause high levels of cholesterol in the blood. Cholesterol deposits contribute to the blockage of arteries.*

malnutrition is the state of poor health caused by a lack of sufficient food, lack of **balanced diet**, or a condition which prevents the body from absorbing or using nutrients properly.

anorexia nervosa is an eating disorder typically involving refusal to eat and an obsessive desire to become slim. *It is becoming increasingly common among girls and young women in affluent countries.*

bulimia is an eating disorder typically involving excessive eating followed by self-induced vomiting.

force (symbol: *F*) A force is a pushing or pulling action which can change the shape of an object, or make a stationary object move or a moving object change its speed or direction. *Forces have direction (**vector quantity**), so they are represented by arrowed lines. The length of the arrow indicates the size of the force and the direction of the arrow its direction. If equal and opposite forces act on an object, the result may be that the object becomes stretched or squashed. The sizes of forces are measured in units called **newtons**.*

TYPES OF FORCES

newton (symbol: N) The newton is the SI unit of **force**, defined as the force which gives a mass of 1 kilogram an acceleration of $1\,\mathrm{ms}^{-2}$. *It was named after the British physicist Sir Isaac Newton (1642–1727).*

magnetic forces and **electric forces** act at a distance due to the force field around a magnet, electric charge, or electric wire (see **magnetic fields** and **electric fields**).

friction (or **frictional force**) is the force which acts to oppose the motion between two surfaces as they move over each other. *It is because of friction that most people think forces are needed to keep something moving (contrary to Newton's first law). Friction helps to give grip and traction between surfaces such as tyres on the road, brakes, etc. Sometimes friction has unwanted effects such as the generation of heat or the wearing of surfaces. Friction can be reduced by:*
- *streamlining which reduces air or water resistance.*
- *lubrication between surfaces with oil or grease.*
- *minimising surface contact by using ball bearings.*

static frictional force (or **limiting frictional force**) is the maximum value of the frictional force between two surfaces which can prevent one surface from sliding over another.

dynamic frictional force is the value of the frictional force when one surface is sliding over another. *This is less than the **static frictional force**.*

gravitational force (or **gravitation** or **gravity**) is the force of attraction that objects have on one another because of their masses. *Like magnetic or electric forces, it is a force that acts at a distance according to the **inverse square law**. Normally the gravitational force between two objects is very weak. However, if one of the objects is massive, such as a planet, the force becomes noticeable. On the surface of the Earth the gravitational force acting on a mass of 1 kilogram is approximately 9.8 N. The size of this* *force becomes smaller as the object moves further away from the surface of the Earth. Outside the Earth's **gravitational field** an object becomes 'weightless'. (Its mass, however, remains the same.) Gravity is the force of attraction which holds comets, moons, satellites and space stations (see **circular motion**) in orbit.*

weight is the **gravitational force** exerted on an object by the Earth (or another planet).

Newton's law of gravitation states that the gravitational force of attraction between two particles is given by the equation:

$$F = \frac{G\, m_1 m_2}{d^2}$$

where F = gravitational force
G = gravitational constant
m_1 and m_2 = mass of two particles
d = distance between the particles

compression forces are equal and opposite forces which when applied to an object result in a decrease in its length.

tension forces are equal and opposite forces which when applied to an object result in an increase in its length.

elasticity is the property of a material of stretching when a force is applied to it and it cannot move. *When the deforming force is removed, the material returns to its original shape provided that its **elastic limit** has not been reached. Rubber, nylon, and coiled springs are elastic materials.*

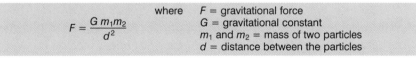

Intermolecular repulsion causes compression forces   Intermolecular attraction causes tension forces

COMPRESSION AND TENSION FORCES

Hooke's law states that the extension of a material is directly proportional to the force that is stretching it. *This law is true only up to the **elastic limit**. It was named after the English scientist Robert Hooke (1635–1703).*

elastic limit The elastic limit is the point beyond which a material loses **elasticity** and stops obeying **Hooke's law**. *After this point, permanent deformation occurs.*

balanced forces are forces acting on an object which remains at rest or travels at a constant speed. *The diagram shows the balanced forces on an aircraft moving at a constant speed (drag and thrust are equal) and constant height (lift and weight are equal).*

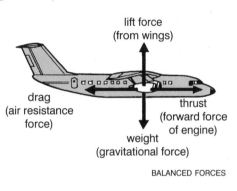

lift force (from wings)

drag (air resistance force)

thrust (forward force of engine)

weight (gravitational force)

BALANCED FORCES

forces (dynamics and turning forces)

dynamics is the study of bodies in motion under the action of forces. *The mass, inertia, and **momentum** of a body all affect the action of a force.*

mass is the quantity of matter in an object (or body). *The unit of mass is the **kilogram**. Mass is also equivalent to **inertia**.*

principle of conservation of mass This principle states that, during any physical or chemical change, mass cannot be created or destroyed. *There is always the same total mass before and after such changes (except nuclear reactions in which mass is converted into energy).*

kilogram (symbol: **kg**) A kilogram is the SI unit of **mass**. *One kilogram is equal to 1000 g, and one metric tonne is equal to 1000 kg.*

inertia is the tendency of an object to resist a change in speed (**acceleration**) caused by a force. *It is directly related to its **mass**. Objects with large mass have a large inertia, and are more difficult to speed up or slow down than smaller objects of low mass and low inertia. Inertia is often described as 'resistance to motion'.*

momentum of an object is its **mass** multiplied by its **velocity**. *Momentum, like velocity, is a **vector quantity** which means that it has both magnitude and direction. An object travelling at constant speed has no net force acting on it, but it does have momentum. The SI unit of momentum is the newton second (N s).*

acceleration is the rate at which the velocity of an object changes.

Newton's laws of motion These are three laws relating forces and motion, formulated by Sir Isaac Newton (1642–1727).

Newton's first law states that an object will continue in a state of rest or uniform motion unless acted upon by an external force. *This law implies that all changes in speed (acceleration) are caused by forces. The relationship between force and acceleration is:*

$$\text{force} = \text{mass} \times \text{acceleration}$$

*An object in free fall is accelerating by gravitational acceleration (g), and the force is **gravitational force** or **weight**.*

$$\text{weight} = \text{mass} \times \text{gravitational acceleration}$$

Newton's second law states that the rate of change of momentum of an object is directly proportional to the force acting on the object. *From this second law we can again deduce the relationship between force and acceleration. Normally the mass of an object is constant, so the force is directly proportional to the acceleration of the object, and acts in the same direction.*

Newton's third law states that forces always occur in equal and opposite pairs called the action and reaction.

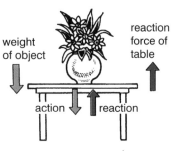

NEWTON'S THIRD LAW

turning force A turning force is a force applied to an object which is fixed at a point around which it may rotate. *The size of the turning force is its **moment**, which can be increased by increasing the force, or increasing the distance from the pivot, or both.*

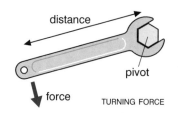

TURNING FORCE

moment A moment of a force is a measure of the ability of a force to rotate an object about a pivot. *The size of the moment is equal to the force multiplied by the perpendicular distance from the axis of the force to the pivot:*

$$\text{moment} = \text{force} \times \text{perpendicular distance to pivot}$$

The SI unit of a moment is the newton metre (Nm).

couple A couple is two parallel **turning forces** which are equal and opposite but which do not act along the same line of action. *The resultant force of a couple is equal to the sum of the **moments**.*

principle of moments This principle states that when an object is in equilibrium (balanced), the sum of the clockwise moments (moments which tend to turn the object in a clockwise direction) is equal to the sum of the anticlockwise moments about the same point.

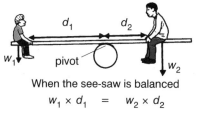

When the see-saw is balanced

$$w_1 \times d_1 = w_2 \times d_2$$

PRINCIPLE OF MOMENTS

equilibrium occurs when the overall clockwise moments acting on an object are equal to the overall anticlockwise moments.

centre of gravity (or **centre of mass**) The centre of gravity is a point on an object through which its total weight (or mass) appears to act. *Objects will balance if supported at their centre of gravity.*

stable equilibrium (or **stability**) is the state of an object which will return to its original position when tilted. *This happens if the centre of gravity is raised when the object is moved. Stability of objects is increased by having a low centre of gravity and a wide base.*

coin lying flat

STABLE EQUILIBRIUM

coin on side

UNSTABLE EQUILIBRIUM

unstable equilibrium is the state of an object which will not return to its original position when tilted. *For such objects, the slightest movement causes the centre of gravity to move to a lower position.*

football on flat surface

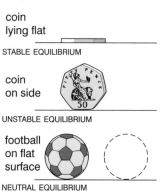

NEUTRAL EQUILIBRIUM

neutral equilibrium is the state of an object which, when moved a small distance from its equilibrium position, remains in the new position. *This happens when the centre of gravity remains at the same height (as with a ball).*

fuels and energy sources

fuel A fuel is a substance which releases useful **heat** or **energy** during combination with oxygen. *Until the 18th century, wood, animal products, and vegetable oil were the main fuels. Since then these have been replaced by fossil fuels.*

fossil fuels are formed from the remains of ancient buried organisms. *Examples are coal, petroleum, and natural gas. All fossil fuels have a high percentage of carbon or hydrogen which, during combustion, combine with oxygen to form carbon dioxide and water. When we burn fossil fuels, we are using the Sun's energy that has been stored as chemical energy underground for millions of years.*

combustion (or **burning**) is a chemical reaction in which a substance (the fuel) reacts rapidly with oxygen and produces heat and light. *All combustion reactions are exothermic. Combustion can happen spontaneously, but usually the substance needs heating first.*

FIRE TRIANGLE

calorific value is a measure of the heat given out per unit mass of fuel during complete combustion. *Calorific values are often used to measure the energy content of foodstuffs as well as fuels (1 calorie = 4.18 joules).*

non-renewable energy source An **energy source** which cannot be replaced once it has been used up. *Fossil fuels are non-renewable energy sources (also called 'non-replenishable' or 'finite' energy sources). It has been estimated that most of the Earth's resources of natural gas and oil will be used up within the next hundred years. Coal may last about 300 years.*

renewable energy source (or **replenishable energy source**) An energy source that can be renewed (e.g. wood, which can be replaced by planting more trees). *Wind, wave, biofuel, and biogas are renewable energy sources and all depend on the Sun's energy. Tidal and hydroelectric energy depend on the force of gravity.*

	Some renewable energy sources	
	Advantages	Disadvantages
Hydroelectric	1 pollution-free 2 immediate response to demand 3 no fuel required	1 big impact on environment 2 initial costs high 3 limited to mountainous areas
Wind	1 pollution-free 2 no fuel required	1 unsightly as need large number of turbines 2 noisy
Geothermal	1 pollution-free 2 no fuel required	1 costs of drilling 2 limited to few places

solar energy is energy from the **Sun**, mainly in the form of light and heat radiation. *The Sun's rays may be used directly to heat water in a solar panel, or focused by a curved mirror to a single point in a solar furnace. Photocells (solar cells) can be used to change sunlight into electrical energy.*

wind power is the use of the motion of the Earth's atmosphere to drive machinery or generators to produce electricity. *Wind has been used as a source of power in windmills, sailing ships, etc. since early times.*

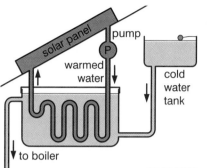

solar panel

pump

P

warmed water

cold water tank

to boiler

SOLAR PANEL

hydroelectricity is electricity produced by trapping rainwater at a high level and then allowing it to flow through electrical turbines at a lower level. *Dams are often built to trap the water high up in the valleys. This water is then released and flows downhill through turbines which drive electrical generators. Gravitational potential energy is changed to kinetic energy, and then to electrical energy.*

tidal energy is produced by the use of tidal barrages to trap water at high tide, which is then allowed to flow through turbines set in a concrete wall. *These turbines drive a generator to produce electricity.*

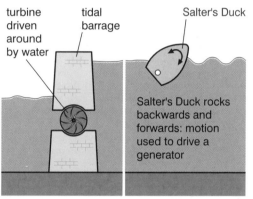

turbine driven around by water

tidal barrage

Salter's Duck

Salter's Duck rocks backwards and forwards: motion used to drive a generator

TIDAL AND WAVE ENERGY

wave energy is the movement of the waves which can be used to rock large floats backwards and forwards. *This movement can then be used to drive a generator to produce electricity.*

geothermal energy is heat energy from hot rock deep in the Earth's crust. *This can be used to heat water in homes, or to make steam for driving turbines to generate electricity.*

biofuel is plant material or animal waste which can be used as a fuel resource. *When it decomposes in the absence of air (anaerobically) it produces **biogas**.*

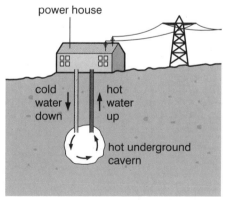

power house

cold water down

hot water up

hot underground cavern

GEOTHERMAL ENERGY

biogas is the gas which is produced from rotting organic matter. *It contains approximately 50% methane and is a useful fuel for heating, cooking, and lighting. Most sewage works and many landfill sites are now designed to collect biogas.*

genetics (inheritance)

genetics is the branch of biology which is concerned with the study of hereditary information transferred from one generation to the next.

gene A gene is the unit of hereditary information, composed of a section of DNA which acts as a chemical instruction for protein synthesis. *Each gene determines the synthesis of a particular protein. Each gene occupies a specific position on a chromosome called its locus. Genes can undergo mutation.*

chromosome A chromosome is a long coil of DNA which is made up of genes in a linear sequence which are found in the nucleus of plant and animal cells. *They are arranged in pairs of homologous chromosomes. Every species has its own number of chromosomes per cell called its diploid number. Humans have 46 (22 matched pairs and one pair of sex chromosomes).*

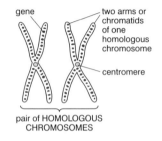

gene
two arms or chromatids of one homologous chromosome
centromere

pair of HOMOLOGOUS CHROMOSOMES

homologous chromosomes are a pair of chromosomes having the same structural features. *Each member of the pair of chromosomes has the same number and pattern of genes, but may have different alleles.*

allele (or **allelomorph**) An allele is an alternative form of a particular **gene**. *In a diploid cell there are usually two alleles of each gene (one from each parent). They occupy the same relative position (locus) on homologous chromosomes.*

nucleic acids are complex organic acids found in a cell's nucleus and responsible for storing and transferring genetic information. *There are two types: DNA and RNA.*

nucleotides are the basic units from which **nucleic acids** are formed.

DNA (abbreviation for **deoxyribonucleic acid**) is a **nucleic acid** which contains the genetic information carried by every cell and directing all the activities of the cell. *It is a large organic molecule with two strands, twisted into a spiral staircase shape (double helix). Each nucleotide is made of a sugar molecule (deoxyribose), a base (adenine A, thymine T,*

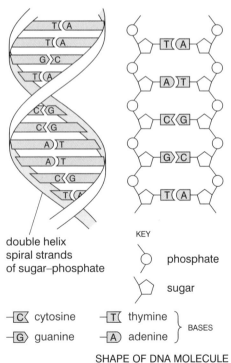

double helix spiral strands of sugar–phosphate

KEY

○ phosphate

⬠ sugar

─C cytosine ─T thymine ⎫
                        ⎬ BASES
─G guanine ─A adenine ⎭

SHAPE OF DNA MOLECULE

guanine G, or cytosine C), and a phosphate. The sequence of nucleotides on one strand of the double helix determines the sequence on the other, as the bases in the nucleotides bind in pairs (A to T and G to C: these are called complementary base pairs).

replication is the production of identical copies of **DNA**. *The two strands separate, and **nucleotides** form a new strand by binding to the correct bases on a piece of single-stranded **DNA**.*

chromatid A chromatid is one of the two 'arms' of a chromosome which has replicated before cell division.

centromere A centromere is the section in the middle of the chromosome where the chromatids join and where there are no genes.

RNA (abbreviation for **ribonucleic acid**) is a single-stranded **nucleic acid**. *Its **nucleotides** differ from DNA in containing ribose (not deoxyribose), and uracil instead of thymine. Messenger RNA takes part in the copying (transcription) of the **genetic code**. Transfer RNA and ribosomal RNA take part in protein synthesis (translation).*

genetic code The genetic code is the code sequence of different bases of a **DNA** molecule, which controls protein synthesis in cells.

haploid describes a cell which has a single set of unpaired chromosomes in its nucleus. *Gamete or sex cells are haploid and are formed by meiosis.*

diploid describes a cell which has paired sets of homologous chromosomes in its nucleus. *In each pair, one chromosome is from the female parent and the other from the male. All cells in animals are diploid (except sex cells) and are formed by mitosis.*

mitosis is division of a cell to form two daughter cells, each with a nucleus containing the same number of chromosomes as the mother cell. *Mitosis is how most cells (except sex cells) divide.*

meiosis (or **reductive cell division**) is division of a cell which results in each daughter cell receiving exactly half the number of chromosomes. *Meiosis takes place in the sex organs to produce **gametes** (sex cells) such as sperm, pollen, and eggs.*

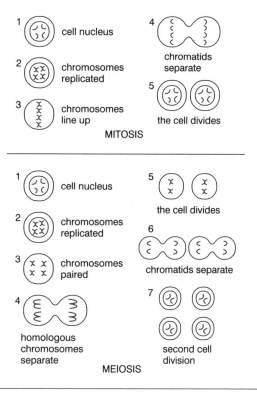

1 cell nucleus
2 chromosomes replicated
3 chromosomes line up
MITOSIS
4 chromatids separate
5 the cell divides

1 cell nucleus
2 chromosomes replicated
3 chromosomes paired
4 homologous chromosomes separate
5 the cell divides
6 chromatids separate
7 second cell division
MEIOSIS

gene pool A gene pool consists of all the genes and their alleles present in a population of a plant or animal species. *No individual in the species can have all the genes. For example, although there are different genes for every hair colour in the UK population, each individual will only have a few of them.*

genome The genome is the entire genetic material of an organism. *In 1988 the international Human Genome Project began to identify all the genes in the 46 chromosomes of every human cell. Computers are used to identify the order of the bases in each DNA strand (DNA sequencing). It is hoped to complete the project by 2010.*

genotype is the genetic information about a particular organism as specified by its **alleles.**

phenotype is the observable characteristics of an organism produced by the interaction of its genes.

dominant allele A dominant allele is a **gene** that will always affect an individual's **phenotype.**

recessive allele A recessive allele is a **gene** that only affects an individual's **phenotype** if it is part of a **homozygous** pair.

Genotype	Phenotype	Description
BB	brown eyes	homozygous
Bb	brown eyes	heterozygous
bb	blue eyes	homozygous

B = dominant allele for brown eyes
b = recessive allele for blue eyes

co-dominance is a situation in a **heterozygous** pair where both alleles are equally dominant. *They are therefore both expressed in the phenotype. For example, the human blood group AB is a result of equally dominant alleles A and B.*

incomplete dominance is a situation in a **heterozygous** pair where neither allele is dominant. *As a result the phenotype is an 'average' of both alleles. For example, the snapdragon plant with alleles for both red and white flowers produces pink flowers.*

Mendel's laws summarise the theory of inheritance. *They were first stated by Gregor Mendel (1822–84), who lived in what is now the Czech Republic.*
- **law of segregation** states that each hereditary characteristic is controlled by two genes (alleles) which during **meiosis** separate and pass into **gamete** or sex cells.
- **law of independent assortment** states that each member of a pair of genes (allele) can join with either of the two members of another pair when the cell divides to form a **gamete** or sex cells.

monohybrid cross A monohybrid cross is the inheritance of a single pair of **alleles.**

F$_1$ (first filial generation) is the first generation of offspring in a genetic cross (in humans, the children).

F$_2$ (second filial generation) is the second generation of offspring in a genetic cross (in humans, the grandchildren).

homozygous describes an organism that possesses identical **alleles** of a particular gene in a given pair of chromosomes. *Two organisms that are* homozygous for a particular character will produce offspring which are also homozygous and identical to the parent with respect to that particular character. *Two organisms that are dominant and recessive homozygous will produce offspring which are all heterozygous..*

heterozygous describes an organism that possesses two different **alleles** of a particular gene in a given pair of chromosomes. *Two organisms which are heterozygous for a particular character will produce equal numbers of offspring which are homozygous or heterozygous with respect to that particular character. The overall phenotype for such a cross is always in a ratio of 3:1.*

variation describes the way in which animals or plants of the same species look or behave slightly differently from each other.

environmental variation is due to **acquired characteristics** such as diet, upbringing, surroundings, etc. *Many characteristics are affected by the environment, but not all. For example, blood group, eye colour, hair colour, and inherited diseases are due to* **genetic variation**.

genetic variation is the variation in the genotype between different individuals. *During* **meiosis** *there is an exchange of genes between* **homologous chromosomes**, *by a process of 'crossing over'. The gametes formed are never exactly the same. When these gametes fuse during* **fertilisation**, *the possible combination of genes which may be produced in the new* **zygote** *is large. Sexual reproduction normally produces new vigour in a species by mixing genetic material to produce genetically varied offspring. It achieves the greatest variation when the parents are drawn from the widest population possible.*

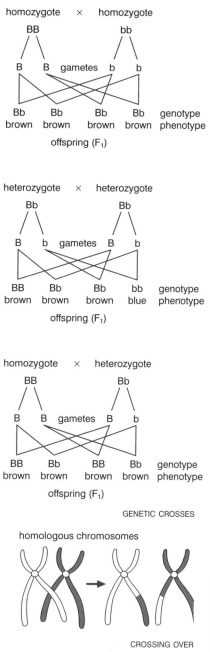

GENETIC CROSSES

CROSSING OVER

genetics (diseases and uses)

sex chromosomes are a pair of chromosomes, found in the nucleus of all cells, which determine sex. *There are two kinds of sex chromosome, the X-chromosome and Y-chromosome.*

	MALE (XY)	
Gametes	X	Y
X	XX	XY
X	XX	XY

FEMALE (XX)

50% Female   50% Male

SEX DETERMINATION

X-chromosomes are the larger of the **sex chromosomes**. A female possesses a pair of these chromosomes (XX).

Y-chromosomes are the smaller of the **sex chromosomes** and cause male characteristics. A male possesses one of each sex chromosome (XY).

sex linkage In human males, a recessive gene carried on the **X-chromosome** will be expressed, because there is no corresponding allele present in the **Y-chromosome** to mask it (as the Y-chromosome is shorter). *In females, the corresponding allele will be present on the other X-chromosome. It is for this reason that females are often carriers of recessive sex-linked disorders, whereas males show the disorder.*

genetic diseases are caused by faulty alleles, often recessive, which can be passed on to future generations.

haemophilia is a **genetic disease** in which a blood clotting substance (factor VIII) is missing or faulty. *The disease is caused by a **recessive allele** carried on the X-chromosome. It therefore affects males, but females can be carriers of the disease.*

Normal male

	Gametes	X	Y
Female carrier	X^h	XX^h	X^hY
	X	XX	XY

Offspring (F₁):
Normal male   XY
Normal female   XX
Carrier female   XX^h
Haemophiliac male X^hY

HAEMOPHILIA

sickle cell anaemia is a **genetic disease** in which abnormality of the haemoglobin causes red blood cells to become crescent-shaped. *These carry little oxygen and get stuck in capillaries, causing blood clots. This disease is caused by a **recessive allele**: if both parents are carriers, there is a 1 in 4 chance for each child that it will develop the disease.*

colour blindness is a **genetic disease** affecting colour vision and results from a malfunction of certain cone cells in the **retina** of the eye.

cystic fibrosis is a **genetic disease** causing an incorrect version of a protein to form in the cell's surface membrane, which results in excess mucus building up in areas such as the lungs. *This disease is caused by a **recessive allele** and is carried by about 1 in 20 people. If both parents are carriers, there is a 1 in 4 chance for each child that it will develop the disease.*

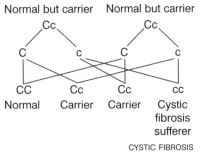

Normal but carrier   Normal but carrier
Cc   Cc
C   c   C   c
CC   Cc   Cc   cc
Normal   Carrier   Carrier   Cystic fibrosis sufferer

CYSTIC FIBROSIS

Down's syndrome results when a person's cells contain three copies of chromosome 21 (instead of a homologous pair). *The problem happens during **meiosis** in the woman's ovaries when one egg cell ends up with both copies of chromosome 21, so that if it is fertilised it will have three copies. The syndrome is characterised by abnormal physical and mental development.*

genetic engineering is a technique of altering an organism's genotype by inserting genes from another organism into its **DNA**. *It can be used to make microorganisms produce useful proteins such as human **insulin**.*

gene therapy is an application of **genetic engineering** which treats disorders (e.g. cystic fibrosis) by the insertion of favourable genes into carriers such as viruses. *Such viruses infect the human cells and reproduce the healthy gene, without causing illness. This is still experimental.*

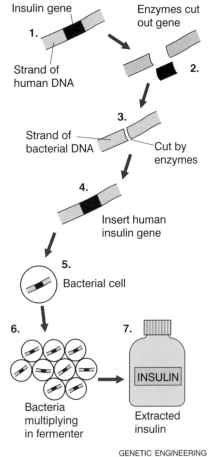

GENETIC ENGINEERING

genetically modified food (abbreviation **GM food**) is food (meat, fruit, cereals, etc.) which has been produced from plants or animals which have had certain genes modified for economic or scientific reasons. *This is another application of **genetic engineering**.*

selective breeding (or **artificial selection**) is the modification of a species by choosing parents (animal or plant) with desirable characteristics. *Such breeding programmes may be used for beef cattle to get the best beef (taste, texture, appearance, etc.). However, the disadvantage is that there is a reduction in the **gene pool**, as the same genes are being chosen each time. It is for this reason that mongrel dogs (random cross-breeds) are often healthier than those chosen by selective breeding.*

genetic fingerprinting (or **DNA fingerprinting**) is a technique in which an individual's DNA is analysed. *This can reveal a sequence of the bases in DNA which is claimed to be specific to the individual concerned. Sufficient DNA can be obtained from very small samples of body tissue such as hair, blood, or semen. The technique is useful in forensic science, paternity disputes, and veterinary science.*

geology is the study of the origin, structure, and composition of rocks.

mineral A mineral is a naturally occurring inorganic substance which has a particular chemical composition and usually a crystalline structure. *Mixtures of minerals together make up a **rock**.*

rock A rock is a mixture of mineral particles making up part of the Earth's crust. *Rocks may be consolidated (e.g. flint, marble, and slate) or unconsolidated (e.g. sand, gravel, and clay). Rocks can be classified into three main groups: **igneous**, **sedimentary**, and **metamorphic**.*

igneous rock is formed when **magma** from the inside of the Earth's crust crystallises. *Such rocks are made up of randomly arranged crystals of a variety of different **minerals**. The size of the crystals in the rock depends on the rate of cooling of the magma.*

intrusive rock is **igneous rock** which has been cooled slowly, usually below the Earth's surface. *The rock only comes to the surface when the overlying rock has been removed by **erosion**. **Granite** is a typical intrusive rock and has large crystals embedded in it.*

intrusion An intrusion is a mass of molten rock which forces its way in between layers of rock. ***Metamorphic** rocks are often formed around an intrusion by the heat and pressure.*

extrusive rock is **igneous rock** which has been cooled quickly on the Earth's surface, perhaps as the **magma** flowed into the sea. *Such rock formed by quick cooling has small crystals embedded in it. **Basalt** is a common, fine-grained extrusive rock.*

sedimentary rock begins to form when existing rock is weathered and the fragments of the rock are deposited, often in the sea or a river. *These rock sediments are then compressed and the water squeezed out, which causes any salts present to crystallise out. These salts then cement the particles together to form the sedimentary rock. Hardened layers of such rock are called strata. The grains of such rock are usually rounded, due to the past effect of water wearing away the jagged edges.*

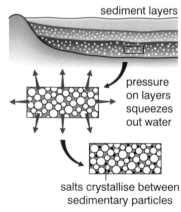

sediment layers

pressure on layers squeezes out water

salts crystallise between sedimentary particles

FORMATION OF SEDIMENTARY ROCK

fossils are remains of dead animals and plants that have been preserved in **sedimentary rocks**. *The organic matter of living material usually rots away, but the hard parts like animal bones or the cellulose and lignin of plant tissue survive for longer. These can slowly absorb minerals from circulating water to replace the decayed organic material, and so the remains are slowly turned into stone (petrification). The oldest fossils are often found buried in the deepest layers of sedimentary rock. More recent*

Type of rock	Type of grain	Mode of formation	Ease of breaking	Fossil remains	Examples
Igneous	grains not lined up	crystallisation of magma	hard, difficult to split	absent	**Basalt**, **Granite**, Pumice
Sedimentary	rounded grains not lined up	deposition of particles	often soft and crumbly	may be present	**Sandstone**, **Limestone**, Mudstone (clay)
Metamorphic	grains often lined up	recrystallisation of other rocks	hard but may split into layers	absent	**Marble** (from limestone), **Slate** (from clay mudstone)

fossils are found in upper layers. However, during movements of the Earth's crust, older rocks and their fossils may be pushed upwards.

metamorphic rock is rock formed by the action of intense heat (from the Earth's mantle) and pressure (from the rocks above) on sedimentary or igneous rock. *Metamorphic rocks are much harder than sedimentary rocks and their grains are usually lined up from the pressure of their formation.*

rock cycle The rock cycle shows how **igneous rock** over millions of years forms **sedimentary rock** or **metamorphic rock** which may eventually melt and resurface as igneous rock once again. *The cycle is driven by **plate tectonic** activity.*

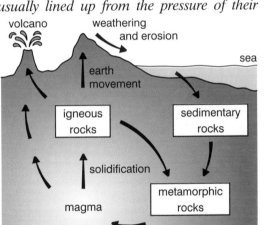

THE ROCK CYCLE

weathering is the wearing down of rocks by the environment.

physical weathering is weathering caused by the sea, wind, ice, temperature changes, etc. *Sea waves continuously pounding the rocks of a cliff, or wind carrying tiny particles of grit, can wear away even the hardest rocks. Ice forming in rock crevices can split rocks, as water expands when it freezes. Great changes in temperature (especially in desert regions) can shatter rock. Plant roots can grow into crevices and break off rock fragments (biological weathering).*

chemical weathering is the dissolving and breakdown of rock by chemical reactions. *Rainwater is a weak acid as it contains dissolved carbon dioxide gas. This can dissolve certain types of rock such as limestone or chalk. Chemical weathering has been increased by the action of **acid rain**.*

erosion is the removal of the weathered parts of rock. *Erosion always involves movement, and rock particles can be carried away by rain, wind, rivers and streams, or the movement of glaciers.*

soil formation occurs when rocks are broken down by physical or chemical weathering. *Physical weathering produces tiny rock fragments of different sizes. Chemical weathering dissolves minerals from the rock fragments to produce clays and soluble mineral salts. Decomposition of organic matter produces* **humus**.

humus is the dark brown organic component of the soil. *It is formed by the action of* **decomposers** *(e.g. bacteria, fungi, earthworms) on dead plants, animal remains, and excrement. Humus lightens the texture of the soil and also provides important nutrients for healthy plant growth. Acidic humus (called mor) is found in coniferous forest, and is formed mainly by fungi. Alkaline humus (called mull) is found in deciduous forest and grassland, and supports an abundance of bacteria and small animals such as earthworms.*

soil profile A soil profile is a vertical section through the soil to show the different layers. *Such layers are sometimes called 'horizons'. For example,* **topsoil** *is horizon A.*

topsoil is the fertile top layer of dark soil, usually less than 20 cm deep. *It is the part of the soil where most of the living organisms exist. Topsoil contains humus, minerals, air, and water.*

subsoil is the lighter coloured layer of soil between the topsoil and bedrock. *It contains coarser rock particles, no humus, and few living organisms. The roots of many trees do penetrate the subsoil as it contains some minerals.*

bedrock (or **parent rock**) is the lowest level of rock, which is usually the original source of the topsoil before weathering.

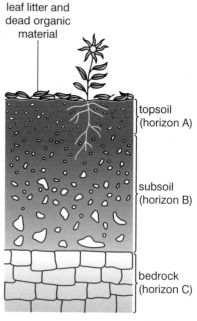

leaf litter and dead organic material

topsoil (horizon A)

subsoil (horizon B)

bedrock (horizon C)

SOIL PROFILE

type of soil The type of soil depends upon the original bedrock it came from, and the amount of humus and minerals it contains. *For healthy plant growth, soils should be rich in* **humus** *and* **minerals**. *The soil should crumble easily so it can trap air, but it should also be able to hold water. Although no two soils are exactly the same we can classify soils into three broad categories according to the average size particle they contain:* **clay soils**, **sandy soils**, *and* **loamy soils**.

clay soil is heavy soil which contains small particles (less than 0.002 mm in size) which can be packed tightly together. *This results in small air spaces between the particles. There is poor drainage in wet conditions, and the soil easily becomes waterlogged. In dry weather it forms cracks. Clay particles have small electrical charges which attract minerals such as* *potassium or calcium ions. Clay soils are therefore rich in minerals, as they are attracted to the soil particles and are not easily leached from the soil.*

sandy soil is light soil which contains large particles (diameters in the range 0.06–2.00 mm) which cannot therefore be packed closely together. *This causes large air spaces between the particles, which results in good drainage.*

loamy soil is a fertile soil made up of a mixture of sand and clay with plenty of organic matter or humus. *The best loamy soils contain 50% clay, 30% sand, and 20% humus. The balance is ideal for plant growth. It will hold water and minerals but will not become waterlogged too easily.*

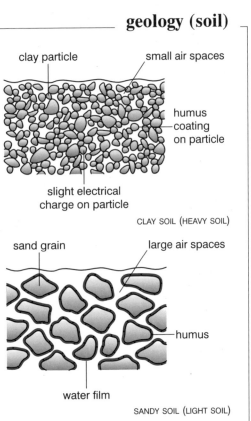

clay particle   small air spaces

humus coating on particle

slight electrical charge on particle

CLAY SOIL (HEAVY SOIL)

sand grain   large air spaces

humus

water film

SANDY SOIL (LIGHT SOIL)

capillarity is the tendency of water to rise into small very narrow spaces. *Clay soils have small air spaces and good capillarity, and dry out slowly. Sandy soils have poor capillarity and dry out quickly.*

alluvial deposits are materials such as soil and silt which have been deposited by rivers. *Alluvial material is often very fertile.*

bacteria in the soil Microscopic bacteria are present in all types of soil. *They are very important as they are **decomposers** and help to break down dead plant and animal remains and excrement. Some are 'nitrifying bacteria' which convert ammonium compounds into nitrates which can be used as nutrients by plants. Others are 'denitrifying bacteria' which release nitrogen back into the atmosphere (see **nitrogen cycle**).*

pH of the soil Particular soils have their own **pH**, depending on the type of parent rock and the amount and condition of the humus. *Waterlogged soils such as peat are acidic as the bacteria cannot get enough oxygen to decompose the plants properly. Clayey soils are often boggy soils with a pH between 4 and 55. Limestone soils are alkaline with a pH of around 8 due to the presence of calcium ions. Most plants grow best in a very slightly acidic pH of 6.5.*

liming soil with a mixture of chalk and lime can help to neutralise excess acidity (e.g. caused by **acid rain**), and can improve drainage in **clay soils** by causing small particles to clump together to form larger particles (flocculation).

geology (plate tectonics)

plate tectonics is the theory that the crust of the Earth is made of rock plates (**tectonic plates**) which rest on the **mantle** and have slowly moved throughout geological time, resulting in **continental drift**.

continental drift is the slow movement of the continents across the surface of the Earth due to the motion of the underlying **tectonic plates**. *This movement is caused by **convection currents** in the semiliquid **mantle**, and occurs at a rate of about 1–2 cm per year. Its existence was first seriously suggested by Alfred Wegener (1880–1930), a German scientist, who observed not only that the eastern coast of South America and the western coast of Africa fitted together, but also that there were similar geological features and matching **fossils** on both continents. It was later found that there are symmetrical patterns of **magnetic stripes** on either side of the Mid-Atlantic Ridge between the two continents, where rock has formed along a **constructive plate boundary**.*

Pangaea was a huge supercontinent which existed for over two hundred million years until the beginning of the Jurassic Period (age of the dinosaurs). *It then broke up into two smaller supercontinents (Gondwanaland and Laurasia) which slowly divided into the present-day continents.*

PANGAEA

tectonic plates are huge sections of the Earth's crust which move across the underlying mantle. *There are six major plates (Eurasian, American, African, Pacific, Indian, and Antarctic) together with a number of smaller ones.*

plate boundaries are areas of volcanic and earthquake activity where **tectonic plates** meet.

- **constructive plate boundaries** (or **divergent plates**) occur where plates move apart from each other and new rocks are produced as convection currents bring molten

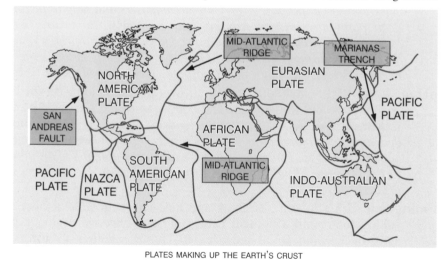

PLATES MAKING UP THE EARTH'S CRUST

magma to the surface. Volcanoes are common at such boundaries. The Mid-Atlantic Ridge is an example of new crust forming at a constructive boundary. Rift valleys may form when plates move apart if a central part slips downward when the plates move apart.

- **destructive plate boundaries** (or **convergent plates**) occur where plates move towards one another so that a more dense plate (oceanic crust) dips below a less dense plate (continental crust) and rejoins the **magma** underneath (**subduction**). This can give rise to deep ocean trenches such as the Marianas Trench in the Pacific. If the two plates have similar densities, neither may be subducted, and they crumple to form fold mountains. An example is the Himalaya range which formed when the Indian and Eurasian plates collided. Mount Everest is getting a few centimetres higher every year as the Indian plate pushes against the Asian plate.
- **conservative plate boundaries** occur where plates slide past one another. The San Andreas fault in California is an example of a conservative boundary. **Earthquakes** are common along such boundaries.

subduction is the dipping of one plate below another at a **destructive plate boundary**.

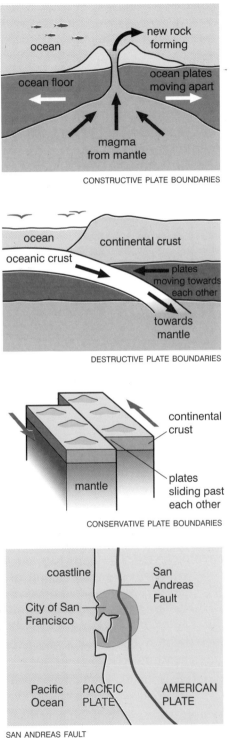

CONSTRUCTIVE PLATE BOUNDARIES

DESTRUCTIVE PLATE BOUNDARIES

CONSERVATIVE PLATE BOUNDARIES

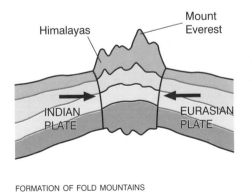

FORMATION OF FOLD MOUNTAINS

SAN ANDREAS FAULT

heart The heart is a muscular organ which pumps blood around the body through the blood vessels. *The heart's pumping action is driven by cardiac muscle in its walls.*

circulatory system The circulatory system consists of the heart, blood, blood vessels, lymphatic vessels, and lymph which together serve to transport materials throughout the body.

double circulation is the type of **circulatory system** found in mammals, with a separate **pulmonary circulation** and **systemic circulation**.

pulmonary circulation is the circulation of **blood** to and from the **lungs**.

systemic circulation is the circulation of **blood** to all parts of the body except the lungs.

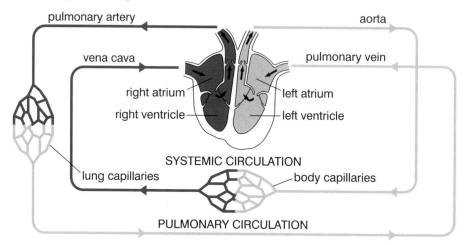

DOUBLE CIRCULATION IN A MAMMAL

atria (singular: **atrium**) (or **auricles**) are the two uppermost chambers of the heart which receive blood from the veins. *They have relatively thin muscular walls and force blood into the ventricles.*

ventricles are the two lowermost chambers of the **heart** which pump blood either to the lungs (right ventricle) or around the body (left ventricle).

cardiac cycle The cardiac cycle is a cycle of events which makes one complete pumping action of the heart. *First both atria contract and pump blood into their respective ventricles, which relax to receive it. Then the atria relax and take in more blood, while the ventricles contract to pump the blood out of the heart.*

heartbeat A heartbeat is the alternate contraction and relaxation of the heart muscles.

diastole is the phase of a **heartbeat** when the heart muscles relax so that the ventricles can fill with blood.

systole is the phase of the **heartbeat** when the heart muscles contract and the ventricles force blood into the arteries.

pulse rate is the number of **heartbeats** per minute (usually around 70).

bicuspid valve (or **mitral valve**) The bicuspid valve in the heart consists of two flaps and prevents blood from flowing back into the left atrium. *It opens to allow blood to flow from the left atrium to the left ventricle.*

tricuspid valve The tricuspid valve in the heart consists of three flaps and prevents blood from flowing back into the right atrium. *It opens to allow blood to flow from the right atrium to the right ventricle.*

semilunar valves are valves with crescent-shaped flaps which prevent backflow, one in the **aorta**, the other in the **pulmonary artery**.

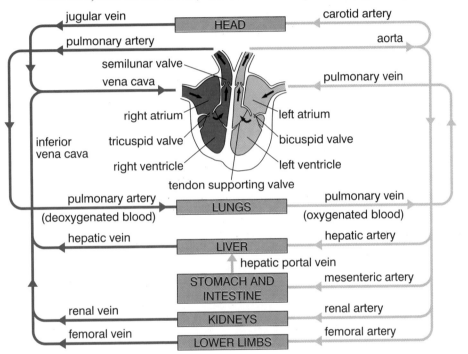

STRUCTURE OF THE HEART AND MAIN ARTERIES AND VEINS

aorta The aorta is the largest **artery** in the body, carrying oxygenated blood out from the left ventricle of the heart.

inferior vena cava The inferior vena cava is a main **vein** which carries deoxygenated blood from the lower body to the heart.

superior vena cava The superior vena cava is a main **vein** which carries deoxygenated blood from the upper body to the heart.

pulmonary artery The pulmonary artery carries deoxygenated blood from the right ventricle to the lungs (the only artery to carry deoxygenated blood).

pulmonary vein The pulmonary vein carries oxygenated blood from the lungs to the left atrium (the only vein to carry oxygenated blood).

heat energy is the energy that flows from one place to another as a result of a difference in temperature. *Energy will always naturally flow from a place of high energy to a place of low energy, and therefore from high temperature to low temperature (except in a **heat pump**).*

thermal energy (or **internal energy**) is the energy an object possesses because of the kinetic and potential energy of its particles. *When an object has heat energy, its particles move and therefore have **kinetic energy**. They also have **potential energy** because their movements keep them separated. When an object absorbs heat energy, its thermal energy increases. When an object loses heat energy, its thermal energy decreases.*

- **heat** and **temperature** are not the same. Heat is the sum of the total energies of all the particles, whereas temperature is a measure of the average energy of the particles. There is more heat in an iceberg than in a cup of boiling water, even though its temperature is lower, because the total energy of the particles in the iceberg is greater.

heat pump A heat pump is a device for transferring heat from a region of low temperature to a region of high temperature by doing **work**, using an energy supply such as an electric motor. *A refrigerator is a heat pump which transfers heat energy from the colder inside to the warmer outside.*

heat capacity (or **thermal capacity**, symbol: *C*) is the heat energy absorbed or released by an object when its temperature changes by 1K. *The heat capacity of an object depends upon its mass and the type of material the object is made of. The SI unit of heat capacity is joule per kelvin (JK^{-1}).*

specific heat capacity (symbol: *c*) is the heat energy absorbed or released when 1 kg of a substance changes its temperature by 1K. *The SI unit of specific heat capacity is joule per kilogram per kelvin ($Jkg^{-1}K^{-1}$). For example, water has a specific heat capacity of $4200\,Jkg^{-1}K^{-1}$. To raise the temperature of 1 kg of water by 1K (1°C) requires 4200J (4.2kJ) to be gained. To raise the temperature of 10 kg of water by 1K requires 42000J (42kJ) to be gained. To lower the temperature of 10 kg of water by 5K requires $42000 \times 5 = 210000J$ (210kJ) to be lost.*

Material	Specific heat capacity ($Jkg^{-1}K^{-1}$)
water	4200
meths	2500
ice	2100
concrete	800
glass	700
steel	500
copper	400
mercury	150

calculations of heat energy Knowing the mass and the specific heat capacity of the substance, you can calculate the amount of energy required to produce a particular temperature rise. *The relationship is given by the following formula:*

Q	$=$	m	$\times$	c	$\times$	ΔT
heat energy (J)		mass in kilograms (kg)		specific heat capacity ($Jkg^{-1}K^{-1}$)		change in temperature (K)

latent heat (symbol: L) is the quantity of heat energy absorbed or released when a substance changes state without changing its temperature. *The heat energy is used to change the arrangement of particles in the various states. For example, when a solid melts to become a liquid, heat energy must be absorbed to move the particles apart. The same is true when a liquid boils to become a gas (or vapour), as heat energy must be absorbed to separate the particles.*

specific latent heat of fusion (symbol: L_F) is the quantity of heat energy absorbed when 1 kg of a substance changes from a solid to a liquid at its melting point. *For water it has a value of 334 000 J kg^{-1}.*

specific latent heat of vaporisation (symbol: L_v) is the quantity of heat energy absorbed when 1 kg of substance changes from a liquid to a gas (or vapour) at its boiling point. *For water it has a value of 2 260 000 J kg^{-1}.*

bimetallic strip A bimetallic strip consists of two metals (usually brass and invar) with different linear expansivities, riveted together. *On heating, the different metals expand by different amounts, which causes the strip to bend. Such strips are used in **thermostats** as they can open and close electric circuits at different temperatures.*

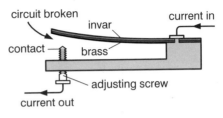

BIMETALLIC STRIP

thermostat A thermostat is a device for automatically controlling the temperature of an appliance such as an oven or central heating system. *The thermostat switches off the current when the appliance is hot enough. It is then switched back on when the temperature falls.*

anomalous expansion of water occurs on cooling water between 4°C to 0°C, as the water expands instead of contracting. *Above 4°C there is normal expansion with increased temperature. This means that water at 4°C has a maximum density. It is for this reason that the surface of a pond will not freeze until all the water is at 4°C. When it does freeze, the warmest water (at 4°C) is at the bottom, and the solid ice, being less dense than the liquid, floats on top. Even when the surface water becomes colder, the denser, warmer water will not circulate. Fish can survive in the warmer water at the bottom of the pond.*

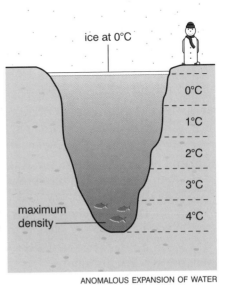

ANOMALOUS EXPANSION OF WATER

temperature is the degree of hotness or coldness of something. *It is a measure of the average **kinetic energy** of its particles.*

thermometer A thermometer is an instrument which is used to measure **temperature**. *There are many types, but all measure how a particular property such as volume, resistance, or e.m.f. changes with temperature.*

liquid-in-glass thermometer The liquid-in-glass thermometer is a common laboratory thermometer which measures temperature by the expansion of a liquid in a very narrow glass capillary tube. *A glass bulb at the bottom acts as a reservoir for the liquid.*

thermometric liquids expand uniformly with temperature and can be used in **liquid - in - glass thermometers**.

clinical thermometer A clinical thermometer is a **liquid-in-glass thermometer** used to measure body temperature. *It has a narow temperature range (35° – 43°C) and a very thin capillary tube to make it very sensitive and precise. The tube has a constriction so that the temperature reading can be taken after the thermometer has been removed from the patient's mouth.*

COMPARISON OF THERMOMETRIC LIQUIDS

	Mercury	Alcohol
For	does not wet sides of tube good conductor so responds quickly	freezes at –115°C expansion greater than mercury
Against	freezes at –39°C	has to be coloured to be visible
	poisonous	thread has tendency to break
	expensive	clings to side of tube

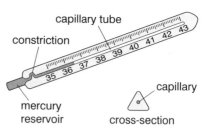

CLINICAL THERMOMETER

capillary tube
constriction
35 36 37 38 39 40 41 42 43
mercury reservoir
capillary
cross-section

maximum-and-minimum thermometer A maximum-and-minimum thermometer is a special **liquid-in-glass thermometer** that records the maximum and minimum temperatures reached during a certain period of time. *A metal index is pushed up or down the tube by the liquid and stays at this maximum or minimum until reset using a magnet.*

press-on thermometer A press-on thermometer is a plastic strip containing 'liquid crystals' which change colour with temperature. *It is placed on the forehead to take the temperature of small children.*

thermocouple thermometer A thermocouple thermometer depends on the temperature difference between two junctions made up of two different metals such as copper and iron. *The greater the temperature difference between the hot and cold junction, the greater the e.m.f. which is produced. Thermocouples are suitable over a wide range of temperatures from about –200°C to 1600°C. Thermocouples can be used in **pyrometers**.*

pyrometer A pyrometer is an instrument used to measure high temperatures above the range of liquid-in-glass thermometers.

fixed points are temperatures which are used as standards for a particular temperature scale. *For example, on the **Celsius scale** the lower fixed point (temperature of pure ice melting) is 0°C and the upper fixed point (temperature of steam at atmospheric pressure) is 100°C.*

fundamental interval The fundamental interval is the distance on the thermometer between the lower and upper fixed points. *On the **Celsius scale** the fundamental interval is divided into 100 degrees.*

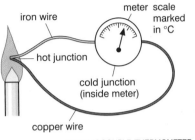

THERMOCOUPLE THERMOMETER

Fahrenheit scale The Fahrenheit scale is an old temperature scale which originally set 0°F at the lowest temperature obtained by mixing ice and salt and 100°F for blood temperature. *This gives values of 32°F for the lower fixed point of ice and 212°F for the upper fixed point of steam. It was named after the German physicist Gabriel Fahrenheit (1686–1736). It is rarely used in scientific work.*

Celsius scale (or **centigrade scale**) The Celsius scale is a common temperature scale based on the lower fixed point of ice at 0°C and the upper fixed point of steam at 100°C. *The graduations on the Celsius scale are identical to those on the **absolute scale** of temperature. It was named after a Swedish astronomer, Anders Celsius (1701–44).*

absolute zero is the lowest temperature theoretically obtainable. *It is 0 K on the **absolute scale**. It is the temperature at which particles have their minimum amount of **thermal energy** and therefore have a minimum temperature and volume. It is equivalent to –273.15°C.*

absolute scale (or **kelvin scale** or **thermodynamic temperature scale**) The absolute scale is based on the lowest possible theoretical temperature of **absolute zero**, which is zero on this scale. *There are therefore no negative temperatures. The relationship between the Celsius and absolute scales is*

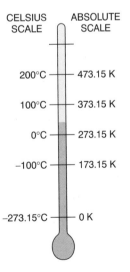

> temperature (K) = temperature (°C) + 273.15

The scale was devised by Lord Kelvin (1824–1907), a Scottish physicist and mathematician.

kelvin (symbol: **K**) is the unit of temperature on the **absolute scale** and is the SI unit of temperature. *The degree sign ° is not used with the kelvin. 1 K represents the same temperature difference as 1°C.*

heat transfer occurs whenever there is a temperature difference as heat energy is transferred from a hotter to a colder place. *This heat transfer continues until both places are at the same* **temperature** *and have the same* **thermal energy**. *There are three possible ways heat energy can be transferred:* **conduction**, **convection**, *and* **radiation**.

conduction (or **thermal conduction**) is the way in which heat energy is transferred through solids (and to a much lesser extent in liquids and gases). *If the solid is a* **conductor** *then heat energy is transferred quickly by movement of free electrons. In* **insulators** *the heat energy is transferred slowly by the vibration of atoms.*

thermal conductivity is a measure of the ability of a substance to conduct heat. *Good conductors have a high thermal conductivity. The SI unit is* $J s^{-1} m^{-1} K^{-1}$.

conductor (or **good conductor**) A conductor is a substance which has a high **thermal conductivity**. *Metals are good conductors because they have lots of free electrons. When a metal is heated, these electrons gain kinetic energy and move more quickly in all directions.*

Conductor	Insulator
silver	sulphur
copper	carbon
aluminium	water
iron	glass
brass (alloy)	plastic
graphite	wood
	rubber
	air

insulator (or **poor conductor**) An insulator is a substance which has a low **thermal conductivity**. *Insulators help to keep hot objects hot and cold objects cold.* **Non-metals**, *wood, plastic, and most liquids and gases are insulators. Materials with trapped air inside, such as expanded polystyrene, wool, fibreglass, etc., are good insulators. The particles in air are far apart, so little heat energy is transferred by collision of particles.*

calorimeter A calorimeter is an apparatus for measuring heat quantities. *Simple calorimeters are made from copper cans.*

convection is the way in which heat energy is transferred through liquids and gases by movement of the particles in the liquid or gas. *If a fluid (liquid or gas) is heated it expands and becomes less dense. Cooler, more dense fluid then sinks, forcing the less dense material upwards against gravity. This circulating movement of a heated fluid is called a* **convection current**. *Natural convection produces ocean currents and creates onshore and offshore winds at the coast during the summer (see diagram).*

radiation (or **thermal radiation** or **radiant heat** or **infra-red radiation**) is the way in which heat energy is transferred from a hotter to a colder place without a medium such as air or water

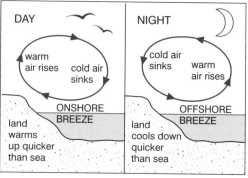

CONVECTION CURRENTS

114

being present. *Thermal radiation can pass through a **vacuum**. The heat energy from the Sun travels to Earth through space as radiant heat. Such heat is an **electromagnetic wave** mainly in the infrared region of the electromagnetic spectrum. When the wave falls on an object, the particles absorb its energy. The particles in the object gain kinetic and potential energy, and the **thermal energy** and **temperature** of the object increases.*

vacuum A vacuum is a space in which there is no matter.

vacuum flask A vacuum flask is a container designed to minimise heat transfer. *It is used to keep hot liquids hot, or cold liquids cold. It is made of a double-walled glass bottle, silvered on the inside and with a **vacuum** between its walls. The bottle has a cork or plastic stopper and is supported by cork blocks in an outer protective case. The vacuum prevents heat loss by **conduction** or **convection** as there is no matter to vibrate or move. The silvered walls of the flask minimise **radiation** loss by reflecting the heat back into the flask.*

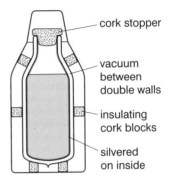

cork stopper

vacuum between double walls

insulating cork blocks

silvered on inside

VACUUM FLASK

absorption of radiation is most efficient with dull (matt) black surfaces; shiny silvery surfaces (which reflect thermal radiation) are weak absorbers.

emission of radiation is most efficient through dull (matt) black surfaces; shiny silvery surfaces are poor emitters. *Heating radiators are misnamed, as they heat mainly by producing convection currents.*

heat exchanger A heat exchanger is a device for removing heat from two fluids (liquids or gases) without allowing the fluids to contact one another. *For example, a car radiator uses a flow of air to cool the hot water in the radiator.*

prevention of **heat loss** in the home and workplace helps to conserve energy.

- loft insulation: fibreglass or vermiculite between ceiling joists reduces **conduction** and **radiation** through the roof
- double glazing: two panes of glass with partial vacuum between reduces **conduction** and **radiation** through windows
- cavity wall insulation: filling the gap between cavity walls with plastic foam reduces **convection** and **radiation** through walls
- carpet or felt on floors: reduces **conduction** and **convection** through floors.

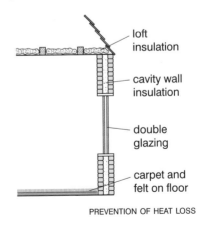

loft insulation

cavity wall insulation

double glazing

carpet and felt on floor

PREVENTION OF HEAT LOSS

homeostasis is the maintenance of a constant internal environment of an organism. *Examples of homeostasis include keeping body temperature constant (in warm-blooded animals), keeping composition of body fluids constant, and maintaining a constant **metabolic rate**. The detection of any deviation from normal is achieved by the passing of information to organs such as the **lungs**, **skin**, **kidneys**, and **liver**. The correction of these deviations is achieved by **negative feedback**.*

metabolic rate is the rate at which an animal uses energy over a given time period. *Metabolic rate is affected by level of activity and temperature.*

metabolism (or **metabolic activity**) is the sum of all the various biochemical reactions that occur in a living organism.

anabolism is the phase of **metabolism** that is concerned with the building up of complicated molecules from simple ones, e.g. protein synthesis.

catabolism is the phase of **metabolism** that is concerned with the breaking down of complicated molecules to simpler ones, e.g. respiration or digestion.

thermoregulation is the control of body temperature in warm-blooded animals. *Heat gain results from metabolic activity. Heat loss occurs mainly through the skin. The **hypothalamus** at the base of the brain continually monitors blood temperature and controls thermoregulation. If the body temperature is too low, it sends instructions (by hormones or through the nervous system) to produce more heat by increasing metabolic activity, and to reduce heat loss through the **skin**. If the body temperature is too high the reverse happens.*

osmoregulation is the control of water content and the concentration of salts in an animal's body. *This is controlled by an area at the base of the brain called the **hypothalamus**, which continually monitors the amount of water in the blood. If the water level is low it sends a message to the **pituitary gland** which secretes the **hormone ADH** (antidiuretic hormone). This hormone makes you feel thirsty, so you drink more water. It also makes the kidneys reabsorb more water, so less is lost by excretion. The reverse also applies: if there is too much water in the blood the pituitary gland stops secreting ADH.*

negative feedback is the process by which information about deviation from a norm passes to a controlling organ and produces a correction of the deviation. *For example, if there is too much glucose in the blood, this is detected by the **pancreas** which secretes a **hormone** called **insulin**. This hormone stimulates the **liver** cells to absorb glucose, which reduces the amount of glucose in the blood, so the pancreas stops secreting insulin.*

lungs The lungs play a major role in controlling the levels of oxygen and carbon dioxide in the blood and body tissue (see **respiration**).

skin The skin and its blood vessels play a major role in controlling the amount of heat which is lost from the body and maintaining a constant body temperature (see **skin**).

kidneys The kidneys control the amount of water and salts in the blood and body tissue *(see **urinary system**)*.

liver The liver is a large and important organ which acts as a 'chemical factory' and has a wide range of functions. *It receives all the digested food dissolved in the blood from the intestines via the hepatic portal vein. It has two main homeostatic functions:*

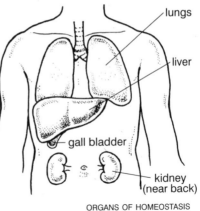

ORGANS OF HOMEOSTASIS

1. To regulate the amount of glucose in the blood. Glucose concentration in the blood is normally around $100\,mg/100\,cm^3$. If there is too much glucose in the blood, the liver removes excess by converting it into **glycogen** *(animal starch), which it stores in its cells. When there is too little glucose, some of the glycogen is broken down to increase blood sugar levels (see **islets of Langerhans**).*

*2. To regulate the amount of amino acids and proteins. Proteins are broken down in the intestines into amino acids, which are then absorbed into the blood and taken to the liver by the hepatic portal vein. Those amino acids that are needed by the body are released back into circulation. Those that are not needed undergo **deamination** to carbohydrate and ammonia. The carbohydrate is stored, but the ammonia is converted to **urea** and excreted via the **kidneys**.*

deamination is the removal of the amino group (-NH$_2$) from an amino acid. *This group is converted to ammonia (NH$_3$) and then combined with carbon dioxide (CO$_2$) to form urea, which is then excreted.*

	Ten important functions of the liver
1	Controls the amount of **glucose** in the blood using the hormones insulin and glycogen.
2	Regulates the amount of amino acids and proteins by converting excess into **urea** and carbohydrates.
3	Makes **bile** which is stored in the gall bladder and helps in the digestion of fats.
4	Stores carbohydrates like glucose as the polysaccharide **glycogen** (animal starch).
5	Stores **vitamins** A and D (both fat-soluble).
6	Makes **cholesterol** which is needed to repair cell membranes.
7	Produces **fibrinogen** which is used by the blood platelets as a clotting substance.
8	Produces heat as a result of the many metabolic reactions.
9	Breaks down old **red cells**, storing the iron and excreting the remaining pigments in bile.
10	Removes harmful substances such as alcohol.

plant hormones are specific chemicals produced by the cells of plants which at very low concentration can affect growth and development. *Such hormones have various uses for both gardeners and farmers.*

- as weedkillers: growth hormones can make weeds grow so fast that they exhaust themselves and die.
- to produce seedless fruit: growth hormones applied to unpollinated flowers cause the fruits to grow without seeds (e.g. seedless grapes).
- as rooting compounds: cut stems can be dipped in rooting hormones to encourage growth.

animal hormones are special chemical 'messengers' secreted in small quantities directly into the bloodstream by an **endocrine gland**. *They travel all over the body, but normally a hormone has a specific effect on one particular organ (target organ) or tissue (target cell). Hormones typically have long-lasting effects, and control things which need constant adjustment or control over long periods.*

endocrine glands (or **ductless glands**) are glands in an animal that secrete hormones directly into the bloodstream. *These hormones dissolve in the plasma and then act at a distant site in the body.*

adrenal glands are **endocrine glands** found above the kidneys which produce **adrenalin**. *This prepare the body for action by increasing the rate of breathing, respiration, and release of glucose from the liver. It also constricts blood vessels in the gut, diverting blood to the brain and muscles where it is needed for action.*

thyroid gland The thyroid gland is an **endocrine gland** at the base of the neck which produces thyroxine under the influence of **thyroid stimulating hormone**, and controls the metabolic rate. *Thyroxine production requires iodine, and iodine deficiency causes the gland to swell and form a goitre.*

islets of Langerhans are a group of specialised cells in the pancreas which secretes two hormones, **insulin** and **glucagon**, to control the level of glucose in the blood.

antagonistic hormones are hormones that produce opposite effects. *Insulin and glucagon are antagonistic hormones.*

diabetes (or **sugar diabetes** or **diabetes mellitus**) is a disorder caused by the lack of the hormone insulin. *This can result in high levels of glucose in the blood (glucose can also be detected in the urine) which can lead to coma and death. Diabetes cannot be cured, but it can be controlled by sensible diet and with daily injections of insulin.*

sex hormones are hormones that control sexual development.

androgens are male sex hormones such as **testosterone**.

oestrogens are female sex hormones such as **oestrogen** itself and **progesterone**.

fertility treatment A hormone called **FSH** (follicle stimulating hormone) can be taken by women which helps to stimulate egg production in the ovaries. *Fertility treatment can sometimes lead to multiple births (twins, etc.).*

contraceptive pill A contraceptive pill contains female **sex hormones** (oestrogens) which, when constantly present in the blood, inhibit the production of **FSH** and stop eggs being released from the **ovaries**.

hormone replacement therapy (abbreviation HRT) is used by women after the **menopause** to replace female sex hormones which would normally be produced during ovulation.

TABLE OF HORMONES

Hormone	Endocrine gland	When secreted	Function
Insulin	islet of Langerhans in pancreas	when blood glucose level rises above normal	stimulates the liver to remove glucose by converting it into glycogen
Glucagon	islet of Langerhans in pancreas	when blood glucose level drops below normal	stimulates the liver to break down glycogen and release glucose back into the blood
Adrenalin	adrenal gland (above kidneys)	small amounts all the time – large amounts when frightened	prepares the body for 'fight or flight'
Thyroxine	thyroid gland (in neck)	throughout life	controls metabolic rate, especially respiration in the mitochondria of cells
Testosterone (androgen)	testes	in larger quantities from puberty onwards, particulary when follicle developing in ovary	controls development of male sex organs and secondary sex characteristics – greater amounts in males gives aggression and a competitive urge
Oestrogen	ovaries	in larger quantities from puberty onwards	controls development of female sex organs and secondary sex characteristics – causes lining of uterus to become thick and spongy
Progesterone	corpus luteum (in ovaries)	after ovulation through pregnancy	maintains the lining of the uterus (lack can cause a miscarriage)
Anti-Diuretic hormone (ADH)	pituitary gland (base of brain)	when quantity of water in blood gets low	allows kidneys to reabsorb water
Thyroid-Stimulating Hormone (TSH)	pituitary gland	throughout life	causes thyroid gland to secrete thyroxine
Growth hormone	pituitary gland	throughout life but especially when young	stimulates growth (lack causes dwarfism, excess causes giantism)
Gonadotrophic hormones	pituitary gland	from puberty onwards	stimulate changes during menstrual cycle
• FSH (Follicle-Stimulating Hormone)			causes an egg to develop in one of the ovaries and stimulates ovaries to produce oestrogen
• LH (Lutenising Hormone)			stimulates the release of an egg at day 14 during the menstrual cycle

male reproductive organ system (or **genital organs** or **genitalia**) comprises two main organs: **penis** and **testes**.

penis A penis is the male reproductive organ of mammals, through which **sperm** (male sex cells) are ejected into the female vagina during copulation. *The **sperm** are produced in the **testes** and travel down sperm ducts and through the **urethra**.*

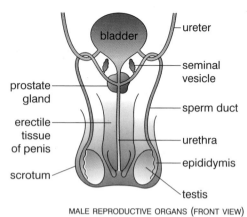

MALE REPRODUCTIVE ORGANS (FRONT VIEW)

*Part of the penis is made of erectile tissue, and it is rich in blood vessels and nerve tissue. When a man is sexually excited it becomes erect and stiff as the blood vessels expand. This facilitates insertion of the penis into the **vagina** ensuring internal fertilisation during **copulation**.*

testes (singular: **testis**) are the pair of male sex organs which are responsible for the production of **sperm** and **androgens** (male hormones). *The testes lie in a sac called the scrotum which hangs outside the body. This allows the testes to be at a temperature slightly below body temperature, which is ideal for sperm production. Sperm are stored temporarily in a coiled tube (epididymis) and are then carried away in the sperm duct (vas deferens) towards the **urethra**. Where the sperm duct joins the urethra there are glands called the **prostate gland** and **seminal vesicle**.*

prostate gland The prostate gland secretes a fluid into the **semen** which activates the sperm and prevents them from sticking together.

seminal vesicle The seminal vesicle secretes a fluid into the **semen** which acts as a source of energy (food) for the sperm.

semen is the fluid from the male reproductive organs which consists of **sperm** from the **testes** and seminal fluid from the **prostate gland** and **seminal vesicle**.

female reproductive organ system (or **genital organs** or **genitalia**) comprises three main organs: **ovaries**, **uterus**, and **vagina**.

ovaries (singular: **ovary**) are the pair of female sex organs which are responsible for the production of **ova** (eggs) and

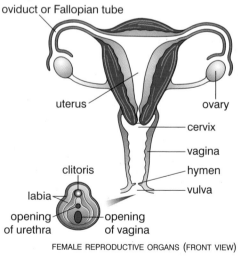

FEMALE REPRODUCTIVE ORGANS (FRONT VIEW)

oestrogen (female hormones). *The ovaries are oval in shape and attached to the back wall of the abdomen, below the kidneys.*

uterus (or **womb**) The uterus is a hollow muscular organ in which a fertilised egg develops into a fetus prior to birth. *The lining of the uterus shows cyclical changes during the **menstrual cycle**.*

ovulation is the periodic release of an **ovum** (egg cell) from the **ovaries** to travel down the oviduct (or Fallopian tube) to the **uterus**, where it is available for **fertilisation**.

vagina The vagina is a muscular tube leading from the uterus whose purpose is to hold the **penis** during mating or **copulation**.

vulva The vulva in female mammals is the external opening of the **vagina**.

clitoris The clitoris is the most sensitive part of the female reproductive organs, and is made of erectile tissue which is rich in blood vessels and nerve endings.

menstrual cycle The menstrual cycle is a roughly monthly cycle of female reproductive physiology during which **ovulation** and **menstruation** occur. *The events of this cycle are controlled by **hormones** produced by the **ovaries** (oestrogen) and **pituitary gland** (gonadotrophic hormones). An area of tissue in the ovary, the follicle, develops to produce an egg. When mature, the follicle bursts, releasing an egg from the ovary (**ovulation**), and changes into the corpus luteum, which secretes **progesterone** hormone, causing the lining of the uterus to become thick and rich in blood vessels. If the egg is fertilised, the developing **embryo** becomes implanted in this lining. If the egg is not fertilised then **menstruation** occurs. The menstrual cycle continues in a woman from **puberty** to the menopause.*

menstruation is the breakdown of the lining of the uterus and its gradual discharge through the vagina (called a period).

menopause The menopause is the time in a woman's life (usually between the ages of 45 and 55) when **ovulation** and **menstruation** stop, and she ceases to be able to have children.

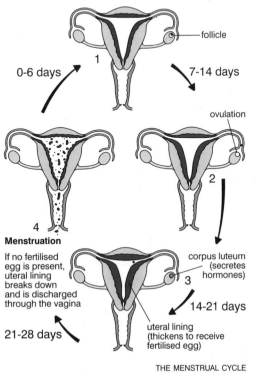

follicle

0-6 days

1

7-14 days

ovulation

2

4

Menstruation

If no fertilised egg is present, uteral lining breaks down and is discharged through the vagina

corpus luteum (secretes hormones)

3

14-21 days

uteral lining (thickens to receive fertilised egg)

21-28 days

THE MENSTRUAL CYCLE

pregnancy (or **gestation**) is the period between implantation of the **embryo** and the birth of the **fetus**. *In humans it lasts about 40 weeks. The implanted embryo in the uterine wall becomes surrounded by amniotic fluid. To begin with the embryo obtains the food and oxygen it needs from blood vessels in the uterus. After a few weeks the placenta and umbilical cord develop.*

amnion (**amniotic sac**) The amnion is the membrane that encloses the **embryo** in the uterus of mammals.

amniotic fluid The amniotic fluid is the watery liquid inside the **amnion** that supports the embryo and protects it from knocks.

amniocentesis is the taking of a sample of **amniotic fluid** from a pregnant woman for microscopic examination of the cells shed from the embryo's skin. *Such examination can determine the condition of the unborn baby, including the presence of Down's syndrome.*

placenta The placenta is an organ in mammals by which the **embryo** is attached to the **uterus** by the **umbilical cord**. *Blood capillaries from the mother and the embryo flow into the placenta and food and oxygen from the mother's blood pass into the embryo's blood by diffusion. Waste material from the embryo also diffuses into the mother's blood.*

umbilical cord The umbilical cord is a flexible tube that connects the **embryo** to the **placenta** in mammals. *Inside this cord is an artery which takes blood from the embryo into the placenta and a vein which returns blood to the embryo.*

embryo An embryo is the organism formed by **mitosis** in the **zygote** so that it is a ball of cells. *It takes several hours for the embryo to reach the uterus, and by this time the ball has 16 or 32 cells.*

fetus (or **foetus**) A fetus is the embryo of a mammal (especially a human) when it has reached a stage of development so that it has recognisable features of the adult form. *In humans the embryo is called a fetus from the eighth week until birth.*

birth (or **parturition**) Birth is the process by which the fully grown fetus separates from its mother at the completion of pregnancy. *The baby is pushed, normally head first,*

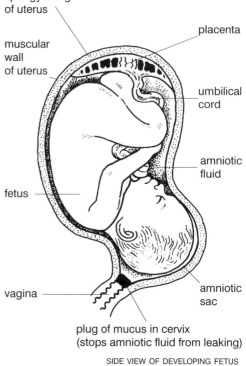

spongy lining of uterus

muscular wall of uterus

fetus

vagina

placenta

umbilical cord

amniotic fluid

amniotic sac

plug of mucus in cervix (stops amniotic fluid from leaking)

SIDE VIEW OF DEVELOPING FETUS

out from the vagina, still joined by the umbilical cord to the placenta. This is cut, and the stump forms the baby's navel.

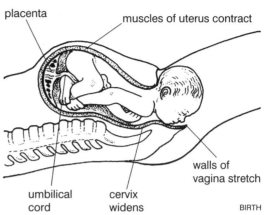

placenta

muscles of uterus contract

walls of vagina stretch

umbilical cord

cervix widens

BIRTH

labour The contractions of the strong muscles of the uterus to push the baby out. *As it passes into the* **cervix**, *the amnion bursts and the fluid escapes (breaking of the waters).*

afterbirth The **placenta**, expelled by further contractions shortly after the birth of the baby.

cervix The cervix is the neck of the uterus at the inner end of the vagina.

identical twins can develop when one egg is fertilised by one sperm, and then splits into two. *Identical twins always share the same* **placenta** *and have the same* **genes** *(same sex and appearance).*

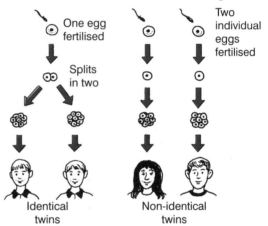

One egg fertilised

Splits in two

Two individual eggs fertilised

Identical twins

Non-identical twins

non-identical twins (or **fraternal twins**) can develop if two or more completely separate eggs are released at the same time by the ovary and separately fertilised. *Each twin has its own* **placenta**.

in vitro fertilisation (from Latin *in vitro* 'in glass') is an artificial method of conception in which the egg is fertilised by a sperm in a laboratory dish or test tube, outside the body of the mother. *The fertilised egg is allowed to divide several times before the embryo is implanted into the mother's womb for a normal pregnancy. The technique is used when women cannot conceive naturally because, for instance, she has blocked Fallopian tubes.*

test-tube baby is an informal term for a baby conceived by in vitro fertilisation. *The first test-tube baby was born in Britain in 1978.*

breech birth A breech birth is when a baby is born bottom first. *In about 4% of births the baby fails to turn around in the uterus to face the* **cervix** *before labour pains begin.*

caesarean birth (or **caesarean section**) A caesarean birth is the surgical removal of the baby from the uterus. *This is carried out with difficult births such as awkward breech births, or when the placenta covers the* **cervix**, *or with large babies and a narrow vagina, or multiple births etc.*

puberty (or **adolescence**) is a stage of development when the reproductive organs begin to function. *It begins approximately between the ages of 11 to 15 in girls and 13 to 15 in boys. It is marked by the start of **menstruation** in females and by the appearance of secondary sex characteristics in both sexes (see table). In males these secondary sex characteristics are controlled by the hormone **testosterone**. In females the characteristics are controlled by the hormone **oestrogen**.*

Female	Male
Growth of pubic hair (around sex organs) and hair under arms.	Growth of pubic hair and other body hair especially facial hair.
Hips become wider (which helps with childbirth later on).	Enlargement of larynx (voice box so voice deepens (voice breaks)).
Sex organs enlarge, ovaries produce eggs (ovulation) and menstrual cycle begins.	Sex organs enlarge and testes produce sperm.
Breasts (mammary glands) develop.	Body becomes more muscular and shoulders broaden.

SECONDARY SEX CHARACTERISTICS

copulation (or **sexual intercourse** or **coitus** or **mating**) is the process by which sperm from a male are inserted into the body of a female. *In mammals the erect penis is inserted into the vagina. Movement of the pelvis stimulates nerve endings in the penis which cause a reflex action called ejaculation. This ejects about 5 cm^3 of **semen** containing around 300 million **sperm** into the top end of the vagina.*

orgasm is the reflex action of ejaculation and the physical excitement associated with it.

conception is fertilisation of the human egg and the implantation in the uterus of the resulting **zygote**.

zygote A zygote is a fertilised egg produced by the fusion of the nucleus of the male and female sex cells. *It is a single cell, but rapidly divides by mitosis to form the **embryo**.*

contraception (or **birth control**) is the avoidance of fertilisation during sexual intercourse. *It is important in limiting the increase in human population. Common methods are shown in the table.*

diaphragm

intra-uterine device

sexually transmitted diseases (abbreviation: **STD** or **venereal diseases**) are diseases that are passed from one individual to another during sexual

condom

pill (oral contraceptive)

METHODS OF CONTRACEPTION

Method	How it works	For and against
Condom (barrier method)	Rubber sheath worn over the erect penis which prevents semen from entering the vagina.	Reliable if used correctly.
Diaphragm (barrier method)	A rubber cap that fits over the cervix and prevents semen entering the uterus.	Reliable if fitted correctly and used with sperm-killing cream around its edges (spermicide).
Spermicide	Jelly or cream which contains chemicals that kill sperm.	Unreliable on their own and should be used in conjunction with condom, diaphragm, etc.
Intra-uterine device (IUD)	Coil or loop inserted into the uterus (womb) by a doctor, which prevents a fertilised egg from implanting in the uterus.	Reliable with most women but can cause heavy periods.
Pill or **oral contraceptive**	Contains chemicals found in pregnant women that prevents ovulation.	Very reliable, but some women experience unpleasant side effects such as nausea, change in breasts, increased chance of blood clots.
Rhythm method	Limits intercourse to the 'safe period', the days in the menstrual cycle when fertilisation cannot take place.	Unreliable, as periods are not always regular and sperm can remain active some time after intercourse.
Sterilisation (**vasectomy** in men, **ligation** in women)	Permanent surgery to cut and tie the sperm ducts or the Fallopian tubes.	Extremely reliable, but normally irreversible, so you cannot change your mind and decide to have children.

METHODS OF CONTRACEPTION

intercourse. *They include bacterial diseases such as **gonorrhoea** and **syphilis,** and viral diseases such as genital herpes and **AIDS**. The transmission of sexually transmitted diseases can be reduced by limiting the number of sexual partners and by the use of condoms. Such devices reduce the risk of contact with body fluids (semen, blood, vaginal fluid) that harbour the microorganisms that cause these diseases.*

gonorrhoea is the disease caused by a bacterium (*Neisseria gonorrhoeae*) transmitted during sexual intercourse. *An infected person may become sterile (unable to produce children). A woman infected with gonorrhoea may pass the disease to her baby at birth. The bacteria in the vagina may enter the baby's eyes and cause blindness.*

syphilis is the disease caused by a bacterium (*Treponema pallidum*) transmitted during sexual intercourse. *The symptoms of the initial stages of the disease are sores and rashes. If untreated, the later stages are deformed joints, paralysis, insanity, and eventual death. As with gonorrhoea, an infected woman can pass the disease to her baby at birth.*

abortion is the expulsion of the fetus from the uterus before the 28th week of pregnancy. *Abortion may be induced or spontaneous.*

miscarriage is spontaneous abortion caused by the rejection of the fetus by the mother's body. *About 25% of pregnancies end in a miscarriage. This may result from poor implantation of the zygote, failure of the placenta to develop, or a deformed embryo.*

hydrogen (symbol: H) is the first and lightest element in the **periodic table**. *It has the simplest atom with a **proton number** of one. Hydrogen is the most abundant element in the universe and is the main constituent of stars. It is present in **water** H_2O and in all **organic compounds**. It has three isotopes, and its overall **relative atomic mass** is 1.008.*

deuterium (symbol: ^2H) is an **isotope** of hydrogen with a neutron as well as a proton in the nucleus of its atom. *It makes up roughly 0.015% of naturally occurring hydrogen.*

tritium (symbol: ^3H) is an **isotope** of hydrogen with two neutrons in the nucleus of its atom. *It is radioactive, and is made artificially.*

hydrogen–1 (symbol: ^1H) is the common **isotope** of hydrogen.

hydrogen bonding is the strong force of attraction between certain molecules that contain hydrogen, such as water molecules. *It is an **intermolecular force** which results when a hydrogen atom is between two electronegative atoms such as oxygen atoms.*

hydrogen ion (chemical formula: H^+) A hydrogen ion is a hydrogen atom which has lost its electron and is therefore just a **proton**. *Hydrogen ions combine with water molecules to form the hydroxonium ion (H_3O^+). Excess hydroxonium ions (hydrated hydrogen ions) make a solution into an **acid**.*

hydrogen molecule (chemical formula: H_2) The hydrogen molecule is the simplest and smallest molecule. *It is a **diatomic molecule** made up of two hydrogen atoms bonded together by a single **covalent bond**.*

hydrogen gas is a colourless, odourless gaseous element made up of hydrogen molecules. *It is the lightest of all gases (density is $0.0899\,g\,dm^{-3}$) and it is difficult to liquefy (b.p. $-253°C$, m.p. $-259°C$). It is a good **reducing agent** and is flammable in air.*

testing for hydrogen When a lighted splint is held at the mouth of a test tube containing hydrogen gas, the gas burns explosively, making a 'popping' sound. *Hydrogen burns in air (or oxygen) with a blue flame to form steam.*

$$\text{hydrogen} + \text{oxygen} \rightarrow \text{water}$$
$$2H_2 \quad + \quad O_2 \quad \rightarrow 2H_2O$$

hydrogen preparation There are three simple ways of making hydrogen gas in the laboratory.

1. *Reaction of reactive metal and water: Sodium and potassium metals react violently with water so it is best to use calcium metal:*

$$\text{calcium} + \text{water} \rightarrow \text{calcium hydroxide} + \text{hydrogen}$$
$$Ca \quad + 2H_2O \rightarrow \quad Ca(OH)_2 \quad + \quad H_2$$

Hydrogen gas is insoluble in water and can be collected using an inverted funnel and test tube full of water.

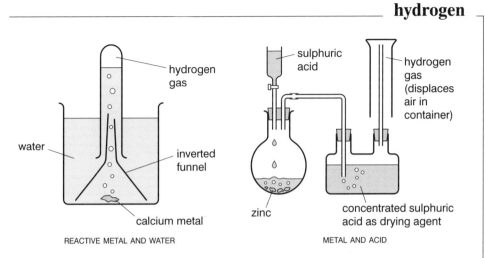

REACTIVE METAL AND WATER   METAL AND ACID

2. **Reaction of metal with acids:** *Suitable metals are iron, zinc, and magnesium. Calcium, sodium, and potassium are much too violent and copper has no reaction with acids.*

> zinc + sulphuric acid → zinc sulphate + hydrogen
> Zn + H_2SO_4 → $ZnSO_4$ + H_2

The hydrogen gas can be dried by bubbling through concentrated sulphuric acid and collected by 'upward delivery and downward displacement of air' in an inverted container.

3. **Electrolysis of acidified water** (*see* **electrolysis**)

> hydrogen ions + electrons → hydrogen gas (collects at anode)
> $2H^+$ + $2e^-$ → H_2

steam reforming is an industrial method of preparation of hydrogen gas. *The raw materials are methane gas (natural gas) and steam, mixed together and passed over a mixture of nickel and iron(III) oxide* **catalysts** *at 1200°C and 50 atmospheres pressure.*

> methane + steam → carbon dioxide + hydrogen
> CH_4 + $2H_2O$ → CO_2 + $4H_2$

uses of hydrogen

- making ammonia by the **Haber process**.
- making **methanol** CH_3OH by reaction with carbon monoxide under pressure over heated catalysts.

> carbon monoxide + hydrogen → methanol
> CO + $2H_2$ → CH_3OH

- **hydrogenation** of oils (addition of hydrogen across unsaturated double bonds) to make saturated fats (e.g. in margarine).
- as a fuel for spacecraft, though there is a risk of explosion and it is hard to liquefy for storage.
- If steam is passed over heated carbon a mixture of carbon monoxide and hydrogen gases are formed. This is a useful fuel called 'water gas'

industrial chemical processes

Haber process This is the process for the manufacture of **ammonia gas** from direct combination of nitrogen and hydrogen gases in the ratio of 1:3 (see equation). *It is immensely important as the ammonia gas is used to manufacture **fertilisers**. It is named after the German chemist Fritz Haber (1886–1934).*

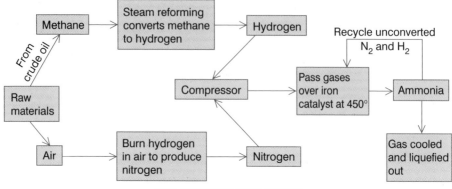

$$\text{nitrogen} + \text{hydrogen} \rightleftharpoons \text{ammonia}$$
$$N_2 + 3H_2 \rightleftharpoons 2NH_3$$

FLOW DIAGRAM FOR HABER PROCESS

conditions for the Haber process reflect the fact that the reaction between hydrogen and nitrogen gases is **reversible** and **exothermic**. *A high yield of ammonia is favoured by low temperature. However, if the temperature is too low the rate of the chemical reaction is too slow. A moderate temperature of 450°C is therefore used, with a **catalyst** of iron and promoters of aluminium and potassium oxides. The higher the pressure, the greater the yield of ammonia. In practice, pressures of around 250 atmospheres are used. Higher pressures pose technical problems and unacceptable expense.*

uses of ammonia

- manufacture of **fertilisers**, especially by reaction with sulphuric and nitric acids to make nitrogenous fertilisers such as ammonium nitrate NH_4NO_3 and ammonium sulphate $(NH_4)_2SO_4$.
- manufacture of **nitric acid** by the catalytic oxidation of ammonia over heated platinum. Uses of nitric acid include making explosives (nitroglycerine, trinitrotoluene (TNT)) and azo dyes (by the reduction of various nitro compounds).
- as a **solvent** in aqueous solution, especially as a degreasing agent since it dissolves grease and fat.

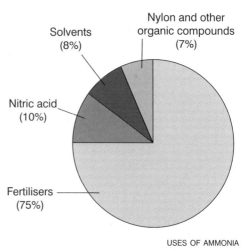

USES OF AMMONIA

contact process This is the process for the manufacture of **sulphuric acid** by the catalytic oxidation of sulphur dioxide to sulphur trioxide.

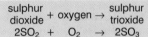

$$\text{sulphur dioxide} + \text{oxygen} \rightarrow \text{sulphur trioxide}$$
$$2SO_2 + O_2 \rightarrow 2SO_3$$

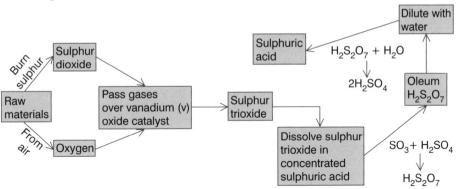

FLOW DIAGRAM FOR CONTACT PROCESS

conditions for the contact process reflect the fact that the reaction between sulphur dioxide and oxygen is **exothermic**, but if the temperature is too low the rate of the reaction is too slow. *A moderate temperature of is 450°C is used together with a vanadium(V) oxide catalyst. The pressure is 2–3 atmospheres and the sulphur trioxide produced (98% conversion) is dissolved in concentrated sulphuric acid to form **oleum**, which is then diluted to form sulphuric acid. The reaction of dissolving sulphur trioxide directly in water is too violent.*

oleum (or **fuming sulphuric acid** or **pyrosulphuric acid** or **disulphuric acid**) is a colourless solid (m.p. 35°C) formed when sulphur trioxide dissolves in concentrated sulphuric acid.

uses of sulphuric acid

- manufacture of **fertilisers** such as ammonium sulphate and calcium superphosphate.
- manufacture of non-soapy **detergents** by sulphonating **organic** molecules with concentrated sulphuric acid.
- as an **electrolyte** in car batteries.
- Other uses include the manufacture of **drugs** (sulphonilamide drugs), paints (lead sulphate), and rayon (artificial silk).

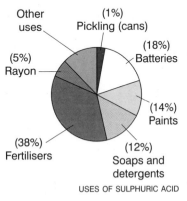

USES OF SULPHURIC ACID

socio-economic factors All industrial processes are influenced by their effects on society, and the economic restrictions needed to make a profit. *These factors may include the cost of raw materials, the site, and the energy needed; the availability of labour, transport and communication facilities; the percentage yields; the risk of pollution and other environmental effects; and the possibility of recycling.*

light (basics)

light is the visible part of the **electromagnetic spectrum** and is a form of **energy** emitted by luminous objects like the Sun. *Light energy travels as waves in straight lines away from its source (rectilinear propagation). The human eye is sensitive to light, and our visual awareness of our surroundings depends upon light.*

luminous objects are objects which emit visible light. *Light is emitted by very hot objects such as the Sun, or the filament in a bulb, or the hot gases in a flame. The intensity of the light emitted by such an object or source is called its luminous intensity and is measured in candelas.*

non-luminous objects are objects which produce no light. *Most objects are non-luminous and can be classified as either **opaque**, **transparent**, or **translucent**.*

opaque describes objects which absorb, scatter, or reflect light and do not allow any light to pass through. *You cannot see through opaque objects.*

translucent describes objects that transmit light but diffuse (scatter) the light as it passes through. *As a result you cannot see clearly through a translucent material such as plastic or obscure glass.*

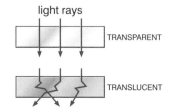

transparent describes objects that transmit light with little or no diffusion or scattering of light. *As a result you can see clearly through a transparent material such as water or glass.*

ray of light A ray of light is a very narrow beam of light energy or radiation. *In diagrams, light is represented by straight lines called rays. An arrow on a ray indicates the direction the light is travelling.*

beam of light A beam is a group of rays of light moving in the same direction. *A beam of light comes from light sources such as projectors, torches, headlights, etc.*

shadow A shadow is an area of darkness on a surface. *It is formed when an **opaque** object prevents light from a source from falling on that surface.*

umbra is the area of total or sharp **shadow** behind an opaque object where no light has reached. *This type of shadow is formed by point sources (see diagram) and has a clearly defined outline.*

penumbra is the area of blurred or fuzzy **shadow** around the

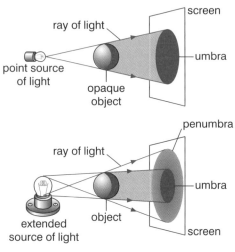

edges of the umbra. *This type of shadow is formed by larger, spread out sources of light. It is an area where a small amount of light has reached.*

eclipse An eclipse is the total or partial blocking of sunlight when one celestial body (Earth or Moon) passes in between the Sun and the other celestial body.

solar eclipse A solar eclipse occurs when the Moon moves into a position directly between the Sun and the Earth. *As a result a circular shadow is cast on the Earth. Such an eclipse appears as*

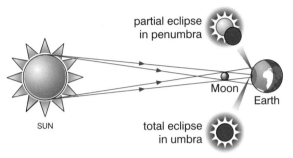

SOLAR ECLIPSE (NOT TO SCALE)

*a total eclipse if it is viewed from the **umbra** region, or as a partial eclipse if viewed from the **penumbra** region. Solar eclipses do not happen very often as, although the Moon circles the Earth every 28 days, only rarely are all three celestial bodies (Sun, Moon and Earth) in a straight line.*

lunar eclipse A lunar eclipse occurs when the Earth moves into a position directly between the Sun and the Moon. *As*

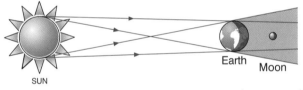

LUNAR ECLIPSE (NOT TO SCALE)

a result an area of full shadow totally covers the Moon. Normally the Moon is out of line with the Earth, and the Sun's light falls on it to give a full moon.

pinhole camera A pinhole camera is the simplest form of camera, consisting of a box with a pinhole at one end and a screen made of tracing paper at the other. *An **inverted** (upside down) **real image** of an object forms on this screen (see diagram). If the pinhole is made larger, the image becomes brighter but also becomes more blurred. This is because a large pinhole acts like lots of small pinholes, each producing an image in a slightly different position.*

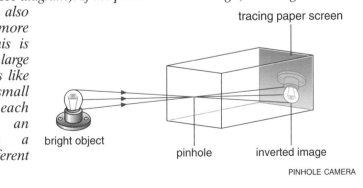

PINHOLE CAMERA

light (reflection and refraction)

reflection of light is the change in direction of a light ray after it hits a surface and bounces off. *Mirrors are normally used to demonstrate reflection because their shiny flat surfaces produce regular reflection. Uneven rough surfaces produce diffuse reflection. Reflected light always obeys the two **laws of reflection**.*

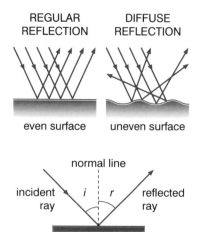

REGULAR REFLECTION   DIFFUSE REFLECTION

even surface   uneven surface

laws of reflection

- The **incident ray**, the **reflected ray** and the **normal line** (at the point of incidence) are all in the same plane.
- The **angle of incidence** is equal to the **angle of reflection**.

normal line   incident ray   *i*   *r*   reflected ray   mirror

> **normal line** A normal line is an imaginary line at right angles to a surface where a light ray strikes it.
> **incident ray** The incident ray is the light ray before reflection.
> **reflected ray** The reflected ray is the light ray after reflection.
> **angle of incidence** (symbol: i) The angle of incidence is the angle between the incident ray and the normal line.
> **angle of reflection** (symbol: r) The angle of reflection is the angle between the reflected ray and the normal line.

image An image is a point from which rays of light entering the eye appear to have originated.

real image A real image is one that can be focused on a screen, as rays of light actually pass through the image. *The image of a **pinhole camera** is real.*

virtual image A virtual image is one which cannot be focused onto a screen because rays of light do not actually pass through the image. *Images formed by a plane mirror are virtual, as only imaginary rays (dotted lines) can be traced back to the image.*

inverted image An inverted image is one that is upside down. *The image of a **pinhole camera** is inverted, and so are the images on the **retina** at the back of the eye.*

lateral inversion is the apparent reversal (left to right, right to left) of the image formed in a plane mirror. *The real reversal is from front to back, as the image is turned through itself to face the object.*

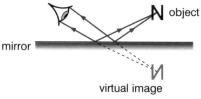

object   mirror   virtual image

LATERAL INVERSION

parallax is the apparent change in the position of an object when it is observed from two different viewpoints. *A fixed object viewed through a mirror appears in different positions when viewed from different angles.*

refraction of light is the change in direction of a light ray as a result of its change in velocity when it passes from one transparent medium (air, glass, water) to another. *If light travels from a less to a more dense* *medium (air to glass) it is slowed down and the light is refracted (bent) towards the normal line in the second medium. Refracted light always obeys the two **laws of refraction**.*

laws of refraction:

- The **incident ray** and **refracted ray** are on opposite sides of the **normal line** and are all in the same plane.
- The value of sin i/sin r is a constant for light passing from one medium to another (i = angle of incidence, r = angle of refraction).

real and apparent depths The **refraction of light** can account for a false impression of depth. *For example objects in water can appear closer than they are (the real depth is greater than the apparent depth).*

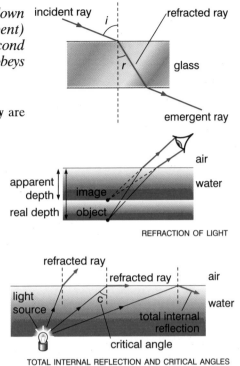

REFRACTION OF LIGHT

TOTAL INTERNAL REFLECTION AND CRITICAL ANGLES

total internal reflection (abbreviation: **TIR**) is the complete reflection of light at a boundary between two media. *The light must travel from more to less dense media and must be incident at the boundary at an angle greater than the **critical angle**.*

critical angle The critical angle is the smallest angle of incidence at which **total internal reflection** occurs (in glass, about 42°; in water, about 45°).

optical fibres use **total internal reflection** to transmit light along very fine tubes of plastic or glass. *Because the fibres are so fine they can be bent without breaking and can carry light around corners. Optical fibres are used in medical viewing instruments and in telecommunications. They are light, relatively cheap, and can carry light messages extremely fast over long distances with little loss in intensity or interference from electrical sources.*

right-angled prisms use **total internal reflection** to turn the path of light through 90° or 180°. *They are used in binoculars, periscopes and cameras.*

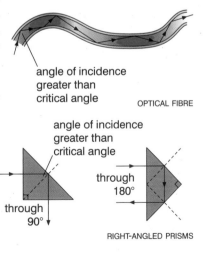

OPTICAL FIBRE

RIGHT-ANGLED PRISMS

lens A lens is a piece of glass which is used to focus or change the direction of a beam of light passing through it. *Lenses are used in a variety of optical instruments such as cameras, projectors, and telescopes.*

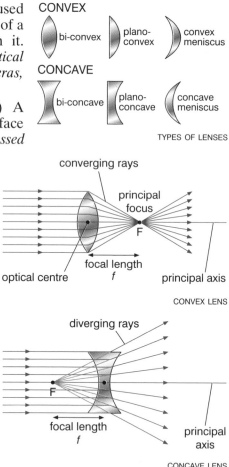

CONVEX

bi-convex — plano-convex — convex meniscus

CONCAVE

bi-concave — plano-concave — concave meniscus

TYPES OF LENSES

convex lens (or **converging lens**) A convex lens is one whose surface curves outwards. *When light is passed through such a lens it is **refracted** towards the **principal axis**. A convex lens is therefore described as a converging lens.*

concave lens (or **diverging lens**) A concave lens is one whose surface curves inwards. *When light is passed through such a lens it is **refracted** outwards away from the **principal axis**. A concave lens is therefore described as a diverging lens.*

converging rays

principal focus

F

focal length
f

optical centre

principal axis

CONVEX LENS

diverging rays

F

focal length
f

principal axis

CONCAVE LENS

optical centre (indicated by black dot) The optical centre is the geometric centre of a **lens**. *Rays of light travelling through the optical centre pass through the lens in a straight line.*

principal axis (or **optical axis**) The principal axis is an imaginary line which passes through the **optical centre** at right angles to the lens.

principal focus (or **focal point**, symbol: F) The principal focus is a point through which all rays travelling parallel to the principal axis pass after refraction through the lens. *A lens has a principal focus on both its sides.*

focal length (symbol: f) is the distance between the **optical centre** and the **principal focus**.

centre of curvature (symbol: C) is the geometric centre of a circle of which the lens surface is a part. *Since a lens has two surfaces, there are two centres of curvature. Notation C is always given to the centre of curvature on the side of the incident light ray (the other is C').*

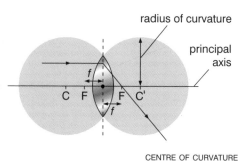

radius of curvature

principal axis

f

C   F   F   C'

f

CENTRE OF CURVATURE

camera A camera contains a **convex lens** which forms a small, **inverted**, **real image** on the photographic film. *When you press the shutter, a hole opens through which light passes onto the film. The size of the hole is called the aperture and is controlled by a diaphragm. The aperture size and the shutter speed (0.1 s to 0.001 s) determine the overall exposure. In dull conditions the shutter speed is longer and aperture larger than in bright conditions. To focus the image, the lens can be moved backwards or forwards. For distant objects, the lens should be the focal length away from the film. For closer objects, the distance between the lens and the film must be greater than the focal length of the lens.*

slide projector A slide projector contains a **convex lens** which forms a magnified, **inverted**, **real image** of a photographic slide (put into the projector upside down) on a screen. *The slide must be positioned between the centre of curvature and the focal point of the convex lens.*

magnifying glass A magnifying glass uses a **convex lens** to produce a magnified, upright, **virtual image** of a small object. *To achieve this, the object must be placed in front of the focal point of the lens.*

astronomical telescope An astronomical telescope uses two **convex lenses** to produce a highly magnified, **inverted**, **virtual image** formed at infinity. *To achieve this, the image of a distant object becomes the object of the second convex lens.*

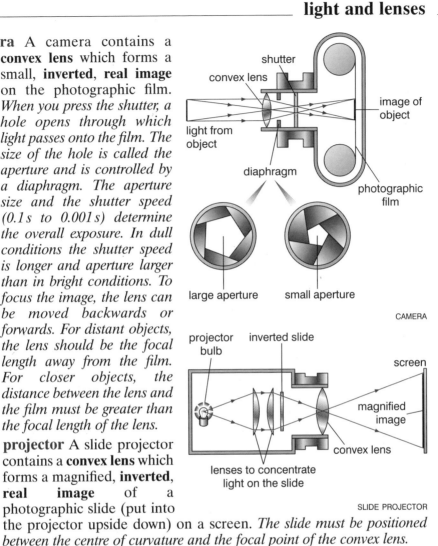

CAMERA

SLIDE PROJECTOR

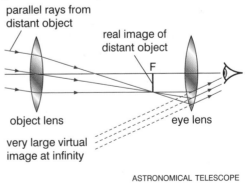

ASTRONOMICAL TELESCOPE

light (colour and dispersion)

colour is the visual sensation that is produced when light of certain **wavelengths** reaches the **retina** of the eye. *The human eye can detect all the spectral colours, each having its own characteristic wavelength of light. Many more colours are possible when mixtures of wavelengths reach the retina. The colour of an object seen by the eye depends on the wavelengths of the light that the object reflects or transmits (other wavelengths being absorbed).*

Spectral colour	Wavelength (nm)
red	620–740
orange	590–620
yellow	570–590
green	500–570
blue	440–500
indigo	370–440
violet	300–370

dispersion is the splitting up of visible light of mixed **wavelengths** by **refraction** into its component wavelengths, which give the different **spectral colours**. *This occurs when visible light passes through a medium such as a glass prism. The shorter wavelengths (blue end of the spectrum) are refracted more than the longer wavelengths (red end of the spectrum), because of the different speeds at which different wavelengths pass through the refracting medium.*

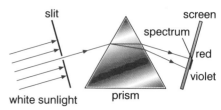

DISPERSION OF LIGHT

spectral colours (or **visible light spectrum**) The seven spectral colours in order of increasing frequency are **red, orange, yellow, green, blue, indigo** and **violet** *(remember by the mnemonic 'ROY G BIV')*.

chromatic aberration is the production of an optical image which has coloured fringes because of **dispersion** at the edges of the lens (see diagram). *It can be corrected by using an achromatic lens.*

achromatic lens An achromatic lens is one that corrects **chromatic aberration** by combining two lenses made of different kinds of glass, such that their **dispersions** cancel each other out but their **refractions** do not.

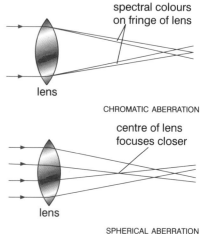

CHROMATIC ABERRATION

spherical aberration is the production of an optical image which is not sharp because the edges of the lens focus in slightly different positions from the centre of the lens.

primary colours with respect to light are any of a set of three coloured lights (**red, blue,** and **green**) which can be

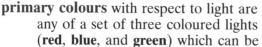

SPHERICAL ABERRATION

mixed together to give white light. *They are called primary because they cannot be made by combining other coloured lights. By adding various primary colour lights in the correct proportions it is possible to produce any spectral colour.*

additive mixing is colour mixing by adding light of different colours (i.e. different wavelengths) to produce an overall colour. *For example, the three primary colours of light will join together to produce an overall white colour. Additive mixing is the method used in colour television.*

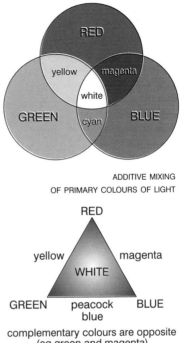

ADDITIVE MIXING
OF PRIMARY COLOURS OF LIGHT

secondary colours with respect to light are colours that can be obtained by **additive mixing** of two primary colours.
- **yellow** is the secondary colour formed by adding red and green light.
- **magenta** is the secondary colour formed by adding blue and red light.
- **cyan** (or **peacock blue**) is the secondary colour formed by adding green and blue light.

complementary colours with respect to light are any two colours which on **additive mixing** produce white light. *For example, yellow and blue are complementary: yellow (a secondary colour) contains two primary colours, and blue is the missing third primary colour.*

RED

yellow   magenta

WHITE

GREEN   peacock   BLUE
blue

complementary colours are opposite
(eg green and magenta)

subtractive mixing is colour mixing by adding together paints, dyes, or pigments, which have colour because they absorb some components of white light and reflect the rest. *It is called subtractive mixing because the colour you see is the colour left after you have subtracted from white light all the wavelengths that have been absorbed. Subtractive mixing is the basis of the process used in colour photography and the mixing of paints.*

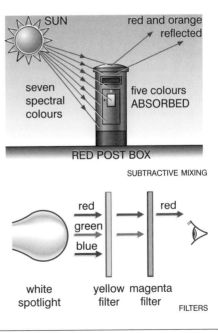

SUBTRACTIVE MIXING

colour filter A colour **filter** absorbs all the light falling on it except for the colour of the filter itself, which it allows to pass through. *For example, a yellow filter will transmit red and green light, and a blue screen will appear black through it. A yellow filter and magenta filter together will transmit red light only.*

137

machines

machine A machine is a device for doing **work**. *In most machines a small force, the effort, is used to overcome a larger force, the load. Such machines are also called force multipliers.*

mechanical advantage (abbreviation: **M. A.**, or **force ratio**) for a simple **machine** is the ratio of the load (output force) to the effort (input force). *A mechanical advantage greater than one means that the load is greater than the effort.*

$$\text{mechanical advantage} = \frac{\text{load}}{\text{effort}}$$

velocity ratio (abbreviation: **V. R.**, or **distance ratio**) for a simple **machine** is the distance moved by the effort (input force) divided by the distance moved by the load (output force) in the same time. *A velocity ratio greater than one means that the effort moves further than the load.*

$$\text{velocity ratio} = \frac{\text{distance moved by effort}}{\text{distance moved by load}}$$

efficiency is the ratio of usable energy output to energy input, expressed as a percentage. *It can also be defined as work or power output divided by work or power input, expressed as a percentage. Efficiency is a measure of how good a machine is at doing its job. A 'perfect machine' would have 100% efficiency, and in such a case the mechanical advantage would equal the velocity ratio. In reality no machine is perfect. Most machines lose useful energy through frictional forces, producing heat and sound.*

Device	Energy input	Energy output	Efficiency
petrol engine	100 J	25 J	25%
diesel engine	100 J	35 J	35%
human	100 J	15 J	15%
power station	100 J	30 J	30%
tungsten filament bulb	60 J	3 J (light)	5%
energy-efficient bulb	15 J	5 J (light)	33%
electric motor	100 J	80 J	80%

lever A lever is a simple **machine** consisting of a rigid bar supported or pivoted at a point along its length called the fulcrum. *An effort applied at one point on the bar can move a load at another point. There are three orders of lever.*

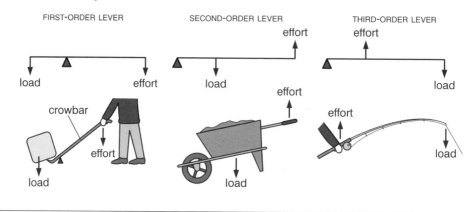

FIRST-ORDER LEVER   SECOND-ORDER LEVER   THIRD-ORDER LEVER

- **first-order lever**: the fulcrum (pivot) is between the load and the effort.
- **second-order lever**: the fulcrum (pivot) is closer to the load than the effort.
- **third-order lever**: the fulcrum (pivot) is closer to the effort than the load.

pulley A pulley is a simple **machine** for raising loads, consisting of one or more wheels with a grooved rim to take a belt, rope or chain. *In the diagram with one pulley (assuming the system is frictionless) the effort force F will lift a load of 2F. The mechanical advantage (and velocity ratio) of this pulley is 2. In the diagram with four pulleys, the effort force F will lift a load of 4F, so the mechanical advantage (and velocity ratio) of this frictionless pulley is 4. This pulley arrangement is called a block and tackle.*

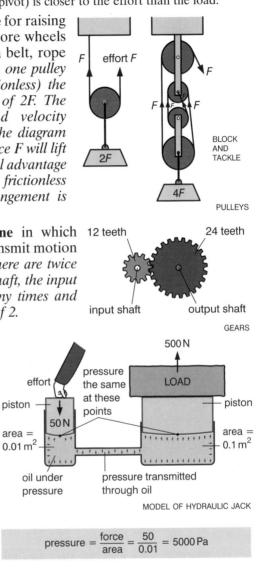

gear A gear is a simple **machine** in which toothed wheels engage to transmit motion between rotating shafts. *If there are twice as many teeth on the output shaft, the input shaft will rotate twice as many times and therefore has a velocity ratio of 2.*

hydraulic machines are **machines** which use liquids under pressure instead of levers or cogs. *They work on the principles that a liquid is virtually incompressible, and that if pressure is applied to the trapped liquid then the pressure is transmitted to all parts of that liquid. The diagram shows a cross-section of a hydraulic jack. If we push down on the smaller piston with a force of 50 N it acts over an area of $0.01\,m^2$.*

$$\text{pressure} = \frac{\text{force}}{\text{area}} = \frac{50}{0.01} = 5000\,\text{Pa}$$

This pressure is transmitted through the liquid, so underneath the larger piston (at the same height) there is exactly the same pressure.

Upward force on large slave piston = pressure × area = 5000 × 0.1 = 500 N

*Therefore, assuming no heat loss from friction, an effort of 50 N pushing down can lift a load of 500 N. Overall the amount of **work** done by both pistons is the same. The smaller piston moves 10 times further than the larger piston as its force is 10 times smaller.*

139

magnetism

magnetism is a property of matter which produces a field of attractive and repulsive forces.

magnetic poles are regions near the ends of a **magnet** from which the magnetic forces appear to originate. *All magnets have equal numbers of poles.*

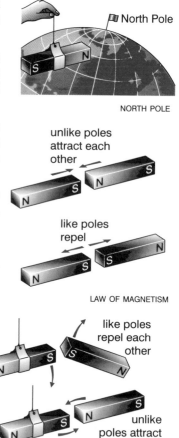

NORTH POLE

north pole (or **north-seeking pole**) A north pole is the end of a magnet which when suspended freely points to the north (magnetic north).

south pole (or **south-seeking pole**) A south pole is the end of a magnet which when suspended freely points to the south.

magnetic axis The magnetic axis is a line joining the north and south poles of a magnet about which the **magnetic field** is symmetrical.

unlike poles attract each other

like poles repel

LAW OF MAGNETISM

law of magnetism states that like poles of two magnets (two north poles or two south poles) repel one another, and unlike poles attract one another. *To find out the polarity of a magnet, bring both its poles in turn near to the known pole of a suspended magnet. If there is repulsion, then the poles are similar in polarity. If there is attraction, then the poles are opposite in polarity.*

like poles repel each other

unlike poles attract each other

TESTING POLARITY

ferromagnetic material (or **magnetic material**) is material which can be magnetised strongly, such as iron, cobalt, nickel, and their **alloys**.

hard magnetic material (or **permanent magnet**) describes a **ferromagnetic material** such as **steel** which retains its magnetism. *Such materials are 'hard' to magnetise, but once magnetised they keep their magnetism.*

soft magnetic material (or **temporary magnet**) describes a **ferromagnetic material** such as pure iron (called 'soft iron') or wrought iron. *Such material is easy to magnetise, but loses most of its magnetism when the external magnetic field is removed.*

non-magnetic material describes materials like silver, gold, copper, brass, aluminium, and non-metals which apparently cannot be magnetised. *However all material has some magnetic property, and very strong magnets can influence non-magnetic material.*

susceptibility is a measure of the ability of a substance to become magnetised. *Magnetic materials have high susceptibility.*

domains are regions in a magnet which, according to the domain theory of magnetism, are made up of many tiny molecular magnets called dipoles. *Within a domain all the dipoles point in the same direction. A magnetic material becomes magnetised when the domains become aligned. When all the domains are aligned the magnet is 'magnetically saturated' (full strength).*

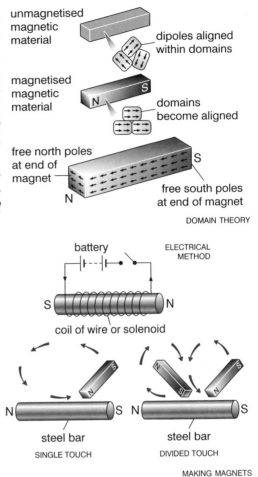

unmagnetised magnetic material

dipoles aligned within domains

magnetised magnetic material

domains become aligned

free north poles at end of magnet

free south poles at end of magnet

DOMAIN THEORY

magnetisation (or **magnetic induction**) is inducing magnetism into magnetic material by aligning its domains. *This only happens when the material is in a magnetic field.*

battery

ELECTRICAL METHOD

coil of wire or solenoid

making magnets by magnetic induction:

- **electrical method** Place the piece of steel in a long coil made of several hundred turns of conducting wire. *Pass a large direct current through the coil for a few seconds. The polarity of the magnet depends upon the direction of the current (see electromagnets).*

steel bar

SINGLE TOUCH

steel bar

DIVIDED TOUCH

MAKING MAGNETS

- **stroking method** The piece of steel is repeatedly stroked with a permanent magnet. *In 'single touch' stroking one pole of a magnet is used. In 'divided touch' stroking the piece of steel is stroked from the centre outwards with two permanent magnets simultaneously. These must have unlike poles (see diagram).*

demagnetisation is the removal of magnetism from a magnetic object by randomising the alignment of the **domains**. *This can be done by hammering the object repeatedly, by strong heating, or by placing it in a changing magnetic field such as that of a coil carrying alternating current.*

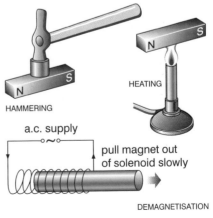

HAMMERING

HEATING

a.c. supply

pull magnet out of solenoid slowly

DEMAGNETISATION

magnetic field A magnetic field is a field of force that exists around a magnet or a current-carrying conductor. *Magnetic objects entering this field are affected by the magnet's forces of attraction and repulsion due to the interaction between their fields.*

magnetic field lines (or **magnetic flux lines** or **lines of magnetic force**) are lines which indicate the direction of the magnetic field around a magnet. *This is the direction a north pole would move in the magnetic field.*

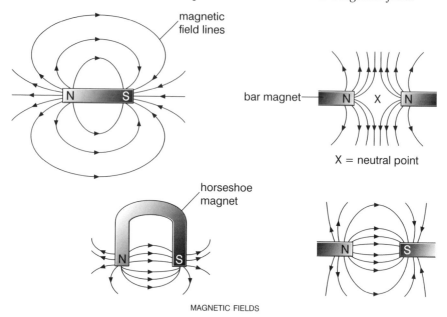

MAGNETIC FIELDS

magnetic flux density is a measurement of the strength of a magnetic field at a particular point. *It is shown by the closeness of the magnetic field lines to each other. Magnetic flux density is always highest around the magnetic poles.*

neutral point A neutral point is a point at which the magnetic flux density is zero (zero magnetism). *Such points occur when equal but opposite magnetic fields cancel each other out, e.g. when two like poles of bar magnets face each other. If a bar magnet is suspended with its north pole facing south, then it has two neutral points in line with its **magnetic axis** (see diagram).*

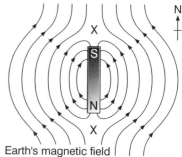

Earth's magnetic field

NEUTRAL POINT

permeability (or **magnetic permeability**) is a measure of the ability of a substance to 'conduct' a magnetic field. *Soft iron has a higher permeability than air, so **magnetic field lines** tend to be concentrated through it (see diagram).*

magnetic shielding is surrounding an object with a material with a high permeability (e.g. soft iron), so that any magnetic field is effectively 'conducted' away from it. *Magnetic shielding is used in sensitive instruments like oscilloscopes.*

the **Earth's magnetic field** is believed to be caused by the oscillating electric fields in the **outer core** of the earth's structure. *The Earth acts as though there were a giant bar magnet through its centre (see diagram). With no other magnets around, a compass needle (which acts as a freely suspended magnet) would point to magnetic north. This differs from true geographic north by the **angle of declination**.*

magnetic stripes are bands of rock of alternate magnetic polarity. *About every half million years, the Earth's magnetic field reverses direction. New rocks take on the new polarity, forming symmetrical stripes on each side of a ridge. These were discovered in the 1960s and provide evidence for **plate tectonics**.*

angle of declination is the angle between true geographic north and a line towards magnetic north. *It varies depending where on the Earth's surface you are, and gradually changes with time, probably because of convection currents in the **outer core** of the Earth. In 1659 the angle of declination was zero and it will be again in about a hundred years.*

angle of inclination (or **angle of dip**) is the angle between a horizontal line and the direction of the Earth's magnetic field at a point on the Earth's surface. *It is measured using a dip circle (see diagram).*

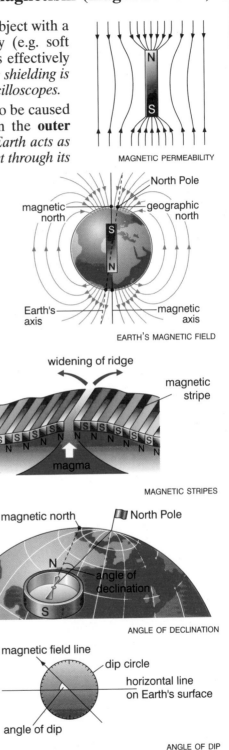

MAGNETIC PERMEABILITY

EARTH'S MAGNETIC FIELD

MAGNETIC STRIPES

ANGLE OF DECLINATION

ANGLE OF DIP

ore An ore is a naturally occurring mineral from which a metal can be extracted on a commercial basis. *Most ores are oxides, carbonates, sulphides, sulphates, or chlorides.*

extraction of metals The method used to extract a metal depends upon its position in the **reactivity series**. *The extraction of a metal from its ore involves the metal ion gaining electrons and is therefore* **reduction**. *Metals which are high in the series form stable compounds, and must be extracted by* **electrolytic reduction**. *Middle order metals can be extracted by heating strongly with a* **reducing agent (smelting)**. *Metals low in the reactivity series form unstable compounds, and can be extracted just by heating the ore.*

Metal	Main ore	Main constituent	Extraction method
potassium sodium calcium magnesium aluminium	carnallite rock salt chalk dolomite bauxite	$KMgCl_3$ NaCl $CaCO_3$ $CaMg(CO_3)_2$ Al_2O_3	electrolytic extraction in which the metal is deposited at the cathode
zinc iron tin lead	zinc blende haematite tinstone galena	ZnS Fe_2O_3 SnO_2 PbS	extraction by heating with coke (carbon) in a furnace
copper mercury	copper pyrites cinnabar	$CuFeS_2$ HgS	roasting by heating the ore

electrolytic reduction is the use of a **cathode** during electrolysis to donate electrons and cause reduction.

aluminium extraction is from its oxide ore **bauxite (Al_2O_3)** by **electrolytic reduction**. *The ore is first purified by dissolving in alkali, and then recrystallised out as pure aluminium oxide (alumina). This is then mixed with cryolite (Na_3AlF_6) which lowers its melting point to about $900°C$. The molten mixture is then electrolysed. Molten aluminium collects at the cathode, and oxygen gas is given off at the anode.*

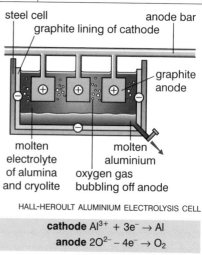

steel cell — anode bar
graphite lining of cathode
graphite anode
molten electrolyte of alumina and cryolite
molten aluminium
oxygen gas bubbling off anode

HALL-HEROULT ALUMINIUM ELECTROLYSIS CELL

cathode $Al^{3+} + 3e^- \rightarrow Al$
anode $2O^{2-} - 4e^- \rightarrow O_2$

smelting is the process of separating a metal from its ore by heating the ore in a furnace with a suitable reducing agent such as coke (carbon). *Usually limestone is added as a fluxing agent to remove impurities such as sand.*

blast furnace (for **iron extraction**) A furnace for **smelting** iron ores such as **haematite (Fe_2O_3)** or **magnetite (Fe_3O_4)** or **siderite ($FeCO_3$)** to make impure iron which is called **pig iron**. *The furnace is charged from the top with the three raw materials: iron ore, coke, and limestone. The furnace is*

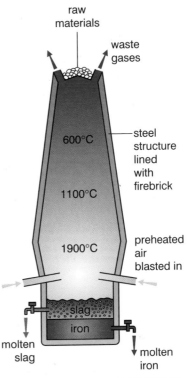

heated by blowing hot air in at the bottom.

1. Coke burns with the oxygen in the air to produce heat and carbon dioxide gas.

$$C + O_2 \rightarrow CO_2$$

2. This carbon dioxide reacts with more coke to give carbon monoxide.

$$CO_2 + C \rightarrow 2CO$$

3. The carbon monoxide reduces the iron ore to molten iron which trickles to the bottom of the furnace.

$$Fe_2O_3 + 3CO \rightarrow 2Fe + 3CO_2$$

4. The limestone in the furnace decomposes to form calcium oxide.

$$CaCO_3 \rightarrow CaO + CO_2$$

5. The calcium oxide is a fluxing agent and combines with impurities such as sand in the ore to form molten calcium silicate or slag. This trickles down the furnace and floats on top of the molten iron.

$$CaO + SiO_2 \rightarrow CaSiO_3$$

BLAST FURNACE

pig iron is impure iron produced in the blast furnace. *It contains about 4% carbon, and is refined to produce* **steel** *or* **cast iron.**

cast iron is hard brittle iron made from remelted **pig iron** mixed with scrap **steel**. *It is good for moulding into complicated shapes such as engine blocks.*

wrought iron is a purer form of iron with 1 to 3% impurities. *It is easy to weld and work (malleable) and is used for chains, hooks, gates, etc.*

steel is an alloy of iron containing small (0.1–1.5%) but controlled quantities of **carbon**. *Most steel is manufactured by the* **basic oxygen process.**

basic oxygen process (or **bop process**) is an industrial process to make steel. *Oxygen gas is blown under high pressure into molten pig iron. This oxidises impurities such as carbon and sulphur to gases which then escape. Other impurities such as phosphorus and silicon are converted to* **acidic oxides**, *which are neutralised by adding a base such as calcium oxide. Very pure iron is then left, to which calculated amounts of* **carbon** *and/or other metals are then added to produce the various steel* **alloys.**

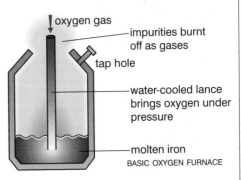

BASIC OXYGEN FURNACE

metal properties Typical metals (excluding Group I and Group II metals) have the following physical properties which make them very useful:

Metallic property	Meaning	Uses
Density (high)	mass per unit volume	lead fishing weights
Durable	resistance to corrosion	zinc dustbins, aluminium windows
Malleable	ability to be made into sheets	cooper roofing, aluminium foil
Ductile	ability to be made into wire	copper wire, iron cables
Sonorous	ability to produce sound when struck	brass (copper alloy) musical instruments, bells
Electrical conductivity (high)	ability to conduct electricity	copper wire, aluminium cables
Thermal conductivity (high)	ability to conduct heat	aluminium saucepans, copper kettles
Tensile strength (high)	strength of metal under stress	steel (iron alloy) bridges

annealing is a type of heat treatment applied to **metals** (ferrous and non-ferrous) to soften them and remove stresses within so they are easier to work or machine, and less likely to shatter. *The treatment consists of heating the metal and then allowing it to cool very slowly.*

alloy An alloy is a **mixture** of two or more **elements** (usually metals except for carbon in steel). *Alloys are often stronger than their constituent elements. Common alloys and their uses are shown in the tables.*

alloys of aluminium are light, fairly strong, and resistant to corrosion as they are protected by a thin oxide coat. *They are used in the aerospace industry, buildings (e.g. window and door frames), overhead power cables, etc. Pure aluminium is a soft light metal, but it can be strengthened by adding small amounts of*

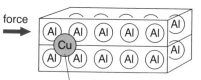

the presence of copper atoms in the alloy makes it harder for the aluminium atoms to slide over one another

ALLOYS OF ALUMINIUM

copper and/or magnesium to make the alloys called **duralumin** *and* **magnalium**. *The new atoms present in these alloys prevent the aluminium atoms from sliding over one another and therefore strengthen its structure.*

uses of metal compounds

sodium chloride (common salt, NaCl) is a common crystalline salt used for seasoning and preserving food (see **salts**).

sodium hydroxide (caustic soda, NaOH) is a white deliquescent (moisture-absorbing) solid which dissolves in water to form the important alkali, aqueous sodium hydroxide. *It is used in the manufacture of* **soaps**, *paper and rayon (see also* **diaphragm cell**).

sodium hydrogen carbonate (bicarbonate of soda, $NaHCO_3$) is a white solid used in baking. *On heating it gives off carbon dioxide gas which makes the dough rise. In solution it is a weak acid, so it is also used as an antacid to relieve acid indigestion.*

sodium carbonate (soda, soda ash, Na_2CO_3) is a white powder which dissolves in water to form an alkaline solution. *In crystalline form (with*

COMMON ALLOYS (NON-FERROUS)

Alloy	Approximate composition	Uses
Cupronickel	75% Cu 25% Ni	'silver' coins
Bronze	90% Cu 10% Sn	medals, swords, statues
Brass	70% Cu 30% Zn	ornaments, electrical contacts
Solder	70% Pb 40% Sn	flux for metals
Pewter	70% Sn 30% Pb	mugs, ornaments
Constantin	60% Cu 40% Ni	thermocouples
Magnalium	70% Al 30% Mg	aircraft frames
Duralumin	95% Al 5% Cu/Mg	construction
Amalgams	Hg/Sn alloys	fillings in teeth

STEEL ALLOYS (FERROUS ALLOYS)

Iron and alloys of iron	Composition	Uses
Wrought iron	99% Fe	garden gates (malleable)
Cast iron	96% Fe 4% C	engine blocks (dense)
Mild steel (low-carbon)	99.5% Fe 0.5% C	car bodies (easily shaped)
High-carbon steel	98.5% Fe 1.5% C	drills (tensile)
Manganese steel	87% Fe 13% Mn	helmets (impact resilient)
Tungsten steel	95% Fe 5% W	cutting tools (very strong)
Stainless steel	18% Cr 8% Ni	cutlery (resists corrosion)

water of crystallization) it is called **washing soda** $Na_2CO_3.10H_2O$. *It is used in the manufacture of **glass** and as a water softener.*

potassium nitrate (saltpetre, KNO_3) is a colourless water-soluble solid which is used in gunpowder, **fertilisers**, and as a preservative in meat.

potassium hydroxide (caustic potash, KOH) is a white deliquescent (moisture-absorbing) solid which dissolves in water to form the alkali aqueous potassium hydroxide. *It is used to make toilet soap.*

magnesium hydroxide ($Mg(OH)_2$) is a white solid which is slightly soluble in water. This suspension (called 'milk of magnesia') is used for acid indigestion and as a laxative (treating constipation).

magnesium sulphate ($MgSO_4$) is a white crystalline solid which is used in medicine as a laxative and in fire-proofing material. *It occurs naturally as hydrated Epsom salts $MgSO_4.7H_2O$.*

calcium oxide (quicklime, lime, CaO) is a white solid which is made in a **lime kiln** by heating calcium carbonate. *It is used to treat acid soils and as a fluxing agent in **blast furnaces**.*

calcium hydroxide (slaked lime, $Ca(OH)_2$) is a white powder which dissolves sparingly in water to form **limewater** $Ca(OH)_2(aq)$. *It is a cheap alkali and is used to treat acid soils, and in the manufacture of whitewash, mortar, bleaching powder, and glass.*

calcium carbonate ($CaCO_3$) is a white solid which is the main constituent of **limestone, chalk,** and **marble**.

calcium sulphate (gypsum $CaSO_4.2H_2O$, plaster of Paris $CaSO_4.\frac{1}{2}H_2O$, anhydrite $CaSO_4$) is a white solid or crystal used in ceramics, plaster, and blackboard chalk.

mixtures and compounds

mixture A mixture is a combination of two or more substances that have not reacted chemically and can be separated using physical processes such as dissolving, crystallization, evaporation, etc. *Examples of mixtures are* **air** *(mixture of gases),* **petroleum** *(mixture of liquids), and* **alloys** *(mixtures of metals). Mixtures are not* **pure substances**.

homogenous mixture A homogenous mixture is a mixture which is the same throughout and therefore has similar properties throughout. *Examples are* **solutions**, *which are homogenous mixtures of solutes and solvents.*

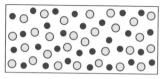

HOMOGENOUS MIXTURE

heterogenous mixture A heterogenous mixture is a mixture with a variable composition and therefore its properties vary from one part to another. *Examples are* **suspensions** *such as chalk in water.*

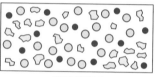

HETEROGENOUS MIXTURE

pure substances contain only one type of atom or molecule. *Elements are pure substances as they contain only one type of* **atom**. **Compounds** *are pure substances as they contain only one type of* **molecule**. *Pure substances have exact melting and boiling points.*

physical change A physical change is one which results in no new chemical substance being formed. *Ice melting or sugar dissolving in ethanol are both physical changes. Such changes are normally easy to reverse. If you cool the water it will change back to ice, and you can separate sugar from ethanol by distillation.*

ice melting

sugar dissolving

PHYSICAL CHANGES

chemical change A chemical change occurs in a **chemical reaction** and produces a new chemical substance. *This substance often looks quite different from the starting substances. For example, when hydrogen burns in oxygen, water is formed. This water is a colourless liquid and has none of the properties of its constituent elements which are both gases. Always during a chemical change there is* **chemical energy**

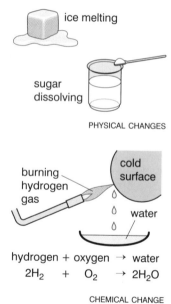

cold surface

burning hydrogen gas

water

hydrogen + oxygen → water
$2H_2$ + O_2 → $2H_2O$

CHEMICAL CHANGE

taken in or given out. When hydrogen burns, heat energy is given out (as chemical bonds are made), as well as light energy (blue flame) and sometimes sound energy (a pop or bang). Most chemical changes are difficult to reverse. Although water can be electrolysed back into hydrogen and oxygen it is difficult to do. Many everyday changes like cooking, rusting, and the decay of food involve chemical changes.

Mixture	Compound
Component substances can be separated by physical means	Constituent elements cannot be separated by physical means
Generally little or no energy is given out or absorbed when formed	Energy is often given out as chemical bonds are being made
Composition can vary	Composition cannot vary
Physical properties (colour, density) are intermediate between those of the substances in the mixture	Physical properties are individual and not a result of its elements
Chemical properties are the result of the substances in the mixture	Chemical properties are quite different from those of its elements

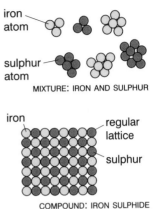

MIXTURE: IRON AND SULPHUR

COMPOUND: IRON SULPHIDE

compound (or **chemical compound**) A compound is the substance formed by the chemical combination of elements in fixed proportions, as represented by the compound's **chemical formula**. *The formation of a compound involves a **chemical change**.*

binary compound A binary compound is made up of two different elements. *Water H_2O, carbon dioxide CO_2, sodium chloride $NaCl$, sulphur dioxide SO_2, ammonia NH_3, and methane CH_4 are all binary compounds.*

synthesis is a **chemical change** by which a compound is built up from its elements or from simpler compounds. *For example, ammonia gas can be built up from its elements (see **Haber process**).*

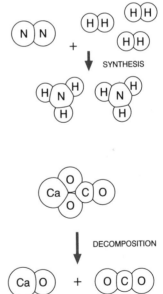

SYNTHESIS

nitrogen + hydrogen → ammonia gas
$$N_2 + 3H_2 \rightarrow 2NH_3$$

decomposition is a **chemical change** by which a compound is broken down into simpler compounds or elements. *Heat is normally required, and such chemical reactions are called **thermal decomposition**. For example, calcium carbonate when heated undergoes thermal decomposition.*

DECOMPOSITION

calcium carbonate	calcium oxide	carbon dioxide

calcium carbonate → calcium oxide + carbon dioxide
$$CaCO_3 \rightarrow CaO + CO_2$$

Compound	Thermal decomposition
Carbonate	metal carbonate → metal oxide + carbon dioxide
Nitrate	metal nitrate → metal oxide + nitrogen dioxide + oxygen
Hydroxide	metal hydroxide → metal oxide + steam

mixtures (solutions and colloids)

solvent A solvent is a liquid that dissolves substances.

solute A solute is the substance which dissolves in the solvent to form a solution. *For example, salt water has water as the **solvent** and common salt as the **solute**.*

solution A solution is a **homogenous mixture** in which the particles of solute and solvent are evenly spread out.

aqueous solution An aqueous solution is one where the **solvent** is **water**. *In chemical equations such solutions are marked (aq).*

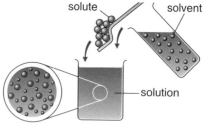

SOLUTION

polar solvent A polar solvent is a liquid (e.g. water) which has polar molecules. *In water, the oxygen atom in the molecule is better at attracting electrons from the covalent bonds than the hydrogen atoms. The oxygen side therefore becomes negatively charged, whereas the hydrogen side becomes positively charged. Polar solvents normally dissolve **ionic compounds** such as common salt.*

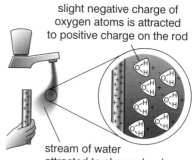

slight negative charge of oxygen atoms is attracted to positive charge on the rod

polarisation is the separation of the positive and negative charge within a molecule (as in a water molecule).

stream of water attracted to charged rod

non-polar solvent A non-polar solvent is a liquid which has non-polar molecules. *Non-polar solvents normally dissolve **covalent compounds** and are often organic liquids such as hexane and tetrachloromethane.*

WATER IS A POLAR SOLVENT

soluble describes a substance (solute) that will dissolve in a liquid (solvent) to form a solution.

insoluble describes a substance that will not dissolve in a liquid.

solubility The solubility of a solute in water, at a particular temperature, is the maximum amount that will dissolve in a given volume of water at that temperature. *Normally the volume of water is 100 cm³ (100 g) so the units of solubility are g/100 g of water.*

saturated solution A saturated solution is one which will not dissolve any more solute at a particular temperature.

solubility curve A solubility curve is a graph to show how the solubility of a solute changes with temperature. *With most solid solutes, the solubility increases with temperature, although with some, like sodium chloride, there is little change. If there is a noticeable increase in solubility with temperature then hot solutions can be made to crystallise out their solute on cooling.*

solubility of gases The solubility of gases normally decreases as the temperature increases. *When a liquid boils all dissolved gases are expelled.*

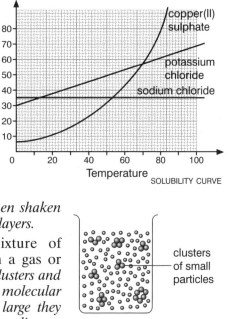

SOLUBILITY CURVE

miscible describes two or more liquids which will diffuse together and form a single **phase** (e.g. alcohol and water).

immiscible describes two or more liquids that will not mix together (e.g. oil and water). *When shaken together such liquids separate into layers.*

suspension A suspension is a mixture of insoluble small solid particles in a gas or liquid. *The particles often stay in clusters and are spread through the liquid by molecular collisions. If the clusters become large they may sink to the bottom and form a sediment.*

clusters of small particles

SUSPENSION

precipitate A precipitate is an insoluble solid formed when a chemical reaction occurs between two dissolved ionic substances. *The clumps of particles in a precipitate are larger than in a suspension and form a sediment.*

phase A phase is any homogenous part of chemical system that is separated from other parts by a definite boundary, e.g. ice mixed with water. *Solids, liquids, and gases are the three phases of matter.*

colloid A colloid is a substance consisting of very small particles (about 10^{-4} to 10^{-6} mm across) suspended and dispersed in a medium such as air or water. *Colloids are intermediate between **solutions** and **suspensions**. They have larger particles than solute molecules in solutions, but smaller particles than in suspensions.*

continuous phase is the phase in a **colloid** throughout which the colloidal particles are dispersed.

disperse phase is the phase in a **colloid** of the colloidal particles themselves.

Type of colloid	Continous phase	Disperse phase	Examples
Aerosol	air	liquid	fog, mist, cloud, paint sprays
Aerosol	air	solid	smoke, dust
Foam	water	gas	whipped cream, fizzy drinks, froth
Gel	water	solid	jelly, gelatin, agar, rubber
Emulsion	water	liquid	milk, salad cream, paint, mayonnaise
Sol	water	solid	paint, milk of magnesia

coagulation is the process by which colloidal particles come together to form large masses which can then be precipitated out.

mixtures (separating techniques)

filtration is a method of separating solid particles from a liquid by passing the mixture through a porous material such as filter paper or glass wool.

residue is the solid trapped in the filter during **filtration**.

filtrate is the clear liquid that passes through the filter during **filtration**.

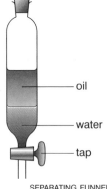

oil

water

tap

SEPARATING FUNNEL

separating funnel A separating funnel is a funnel with a tap used to separate **immiscible** liquids such as oil and water. *The lighter liquid (oil) collects above the heavier liquid (water). When the tap is opened the water is run out, but the tap is closed before the oil reaches the bottom. Separation is never entirely complete.*

decantation is the process of carefully pouring away the liquid above a precipitate or suspension once it has settled. *You can also decant **immiscible** liquids.*

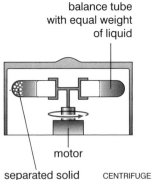

balance tube with equal weight of liquid

motor

separated solid   CENTRIFUGE

centrifuge A centrifuge is an apparatus for the separation of substances by rotating them in a tube in a horizontal circle at high speed. *It can be used to separate fine insoluble particles in a liquid **suspension** or denser liquids from less dense ones. The denser particles are flung to the bottom of the test tube. The lighter particles can be decanted off.*

direct evaporation is used to separate a dissolved solid from a solution by heating. *The **solvent** evaporates and the **solute** is left behind.*

steam evaporation is slow evaporation by heating the solution using a water bath. *It is useful if the **solvent** is flammable and also prevents loss of **solute** due to spitting.*

crystallisation is the process of forming crystals by heating a solution to evaporate some of the solvent. *The hot concentrated solution is then allowed to cool and crystals appear. Crystallisation will only work if a solute is more soluble in hot water than cold.*

simple distillation is the process of boiling a liquid and then condensing the vapour. *Simple distillation is used to purify liquids or to separate a pure liquid from a solution. Normally the condenser used is a **Liebig condenser**.*

Liebig condenser A Liebig condenser is a straight glass tube surrounded by a glass jacket through which cold cooling water is circulated. *It is named after the German chemist Justus von Liebig (1803–73).*

distillate A distillate is the condensed liquid obtained by **distillation**.

distilled water is water purified by **distillation** so that it is free from any dissolved salts or gases.

fractional distillation (or **fractionation**) is a method of separating mixtures of liquids by distillation. *These liquids must be **miscible** and have different boiling points. Separation is achieved by using a long vertical **fractionating column** attached to the distillation flask. Consider separating a mixture of water (boiling*

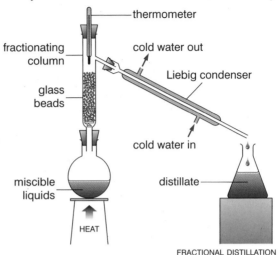

FRACTIONAL DISTILLATION

*point 100°C) and ethanol (boiling point 78°C). When the temperature at the top of the fractionating column reaches 78°C, molecules of ethanol can remain as vapour and pass over into the Liebig condenser. Water molecules, with a higher boiling point, condense and fall back into the flask. This continues until most of the ethanol has boiled off. Then the temperature rises to 100°C and the water vapour passes into the condenser. Fractional distillation is important in the separation of the components of **air** and **petroleum**.*

fractionating column A fractionating column is a long vertical tube packed with glass beads or some other unreactive substance. *This provides a large surface area for condensation and re-evaporation.*

chromatography is a method of separating a mixture by carrying it in solution (or in a gas stream) across an absorbent material. *Chromatography can be used to separate mixtures of gases, liquids, or dissolved substances. It is a simple technique but it is very sensitive. Methods similar to this are important for identifying molecules for medical and biochemical analysis.*

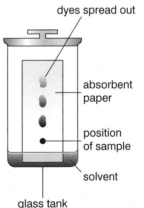

PAPER CHROMATOGRAPHY

paper chromatography is a method of separating dissolved substances, such as dyes and pigments, by spreading them over absorbent paper (e.g. filter paper) with a suitable solvent. *Solutes which are more soluble in the water trapped within the paper fibres travel less far in the solvent.*

chromatogram A chromatogram is the result obtained by chromatographic separation.

moles (chemical calculations)

mole (symbol: **mol**) The mole is the SI unit of 'amount of substance'. *This 'amount of substance' is defined as that which contains the **Avogadro constant** of particles (atom, ions, or molecules).*

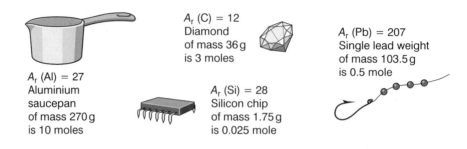

A_r (C) = 12
Diamond
of mass 36 g
is 3 moles

A_r (Pb) = 207
Single lead weight
of mass 103.5 g
is 0.5 mole

A_r (Al) = 27
Aluminium
saucepan
of mass 270 g
is 10 moles

A_r (Si) = 28
Silicon chip
of mass 1.75 g
is 0.025 mole

Avogadro constant (or **Avogadro number**, symbol: L) The Avogadro constant is the number of particles in one **mole** of substance. *It has the value of 6.02×10^{23} as defined by the number of atoms in 12g of the carbon-12 isotope. If you multiply one unit of relative atomic mass by the Avogadro constant you convert it to grams. The constant is named after the Italian scientist Count Amedeo Avogadro (1776–1856).*

molar mass (or **formula mass**) is the mass of 1 mole of particles. *For **elements** this is the mass of 1 mole of atoms and is equivalent to the relative atomic mass of the element in grams. For **compounds** this is the mass of 1 mole of molecules and is equivalent to the relative molecular mass in grams.*

molar volume (or **gram molecular volume** or **molecular volume**) is the volume occupied by 1 mole of any gas at a particular temperature and pressure. *Normally molar volume is measured at room temperature and pressure (r.t.p.) and occupies $24\,000\,cm^3$ ($24\,dm^3$ or $0.024\,m^3$).*

Gay-Lussac's law states that when gases react chemically their volumes are in a simple ratio to one another and to the volume of the products, providing the volumes are measured at the same temperature and pressure. *The law was first stated in 1808 by the French chemist Joseph Gay-Lussac (1778–1850) and led to **Avogadro's law**.*

N_2	+	$3H_2$	→	$2NH_3$
1 vol.		3 vol.	→	2 vol.
$25\,cm^3$		$75\,cm^3$		$50\,cm^3$

EXAMPLE OF GAY-LUSSAC'S LAW

Avogadro's law (or **Avogadro's hypothesis**) states that equal volumes of all gases contain the same number of particles at the same temperature and pressure. *This means that one mole of any gas occupies a certain volume called the **molar volume**. The law is only true for **ideal gases**.*

molar solution (or **molarity**, symbol: **M**) A molar solution contains 1 mole of solute in 1 dm^3 (1 litre) of solution. *It is an indication of concentration of solution. Molar solutions have units of $mol\,dm^{-3}$. A 1 M solution contains 1 mole of solute in $1\,dm^3$ of solution (1 $mol\,dm^{-3}$). Dilute bench acids or alkalis are around $2\,mol\,dm^{-3}$ and concentrated acids are around $10\,mol\,dm^{-3}$.*

moles (chemical calculations)

Type of calculation	$A_r(H) = 1$, $A_r(C) = 12$, $A_r(O) = 16$, $A_r(F) = 19$, $A_r(Na) = 23$, $A_r(S) = 32$, $A_r(Ca) = 40$
GRAMS TO MOLES	How many moles in 88 g of carbon dioxide, CO_2? **number of moles** $= \dfrac{\textbf{mass in grams}}{\textbf{mass of 1 mole}} = \dfrac{88}{44} = 2$ **moles**
MOLES TO GRAMS	How many grams in 0.25 mole of oxygen molecules, O_2? **mass in grams = number of moles × mass of 1 mole** $= 0.25 \times 32 = 4\,g$
FINDING A FORMULA FROM MASSES	20 g of calcium reacts with 19 g of fluorine to form calcium fluoride What is its chemical formula? moles of calcium = 0.5 mole $\qquad Ca_{0.5}F_{1.0}$ moles of fluorine = 1.0 mole $\qquad$ Formula CaF_2
MOLES TO PARTICLES	How many particles in 11 g of carbon dioxide, CO_2? **number of = number of × Avogadro** **particles $\quad$ moles $\qquad$ constant** $= \dfrac{11}{44} \times 6.02 \times 10^{23} = 1.50 \times 10^{23}$ particles
VOLUME OF GAS TO MOLES OF GAS	How many moles in 12 dm^3 of oxygen gas O_2 at room temperature and pressure? **number of moles** $= \dfrac{\textbf{volume}}{\textbf{molar volume}}$ $= \dfrac{12}{24} = 0.5$ mole
MOLES OF GAS TO VOLUME OF GAS	What volume (at room temperature and pressure) would 16 g of oxygen gas O_2 occupy? **volume of gas = number of moles × molar volume** $= \dfrac{16}{32} \times 24 = 12\,dm^3$
CONCENTRATION IN MOLES TO CONCENTRATION IN GRAMS	What is the concentration of a solution of sulphuric acid, H_2SO_4 in g dm^{-3} if it contains 2 mol dm^{-1}? **concentration = concentration × M_r of** **(g dm^{-3}) $\qquad$ (mol dm^{-3}) $\qquad$ solute** $= 2 \times 98 = 196\,g\,dm^{-3}$
FINDING THE CONCENTRATION OF A SOLUTION	What is the concentration (mol dm^{-3}) of a solution if 20 g of sodium hydroxide NaOH is dissolved in 250 cm^3 of water? $\dfrac{\textbf{concentration}}{\textbf{(mol dm}^{-3}\textbf{)}} = \dfrac{\textbf{number of}}{\textbf{moles}} \times \dfrac{\textbf{1000}}{\textbf{volume in cm}^3}$ $= \dfrac{20}{40} \times \dfrac{1000}{250} = 2\,mol\,dm^{-3}$
FINDING THE NUMBER OF MOLES IN A CERTAIN VOLUME OF SOLUTION	How many moles are present in 100 cm^3 of 2 mol dm^{-3} sodium hydroxide solution? **number of moles** $= \dfrac{\textbf{concentration}}{\textbf{(mol dm}^{-3}\textbf{)}} \times \dfrac{\textbf{volume in cm}^3}{\textbf{1000}}$ $= 2 \times \dfrac{100}{1000} = 0.2$ mole
FINDING THE VOLUME OF SOLUTION WHICH CONTAINS A CERTAIN NUMBER OF MOLES	What volume of 2 mol dm^{-3} sodium hydroxide solution contains 0.5 mole? **volume in cm^3** $= \dfrac{\textbf{number of moles}}{\textbf{concentration (mol dm}^{-3}\textbf{)}} \times 1000$ $= \dfrac{0.5}{2.0} \times 1000 = 250\,cm^3$

motion (speed and acceleration)

motion is change in an object's position. *When a **force** acts on an object which is free to move, it may cause the object to move. The direction of movement is in the direction the force is acting, and a single force causes linear (rectilinear) motion.*

displacement is the distance and direction an object has moved from a fixed reference point. *Displacement is a **vector quantity** as it has both size and direction. (Note: distance is a **scalar quantity** as it has only size and no direction.)*

speed is the rate at which an object moves, expressed as the distance the object travels in a certain time. *If the speed of an object remains the same, it is moving with constant or uniform speed. Speed is a **scalar quantity** as it has size but no direction. Speed is often scientifically expressed in units of metres per second ($m\,s^{-1}$).*

velocity is the rate at which an object moves in a particular direction expressed as the displacement of an object in a certain time. *Unlike speed, velocity is a **vector quantity** as it has both size and direction. However, it is measured in the same units.*

stopping distance The stopping distance of a car is the distance it travels from the time the driver decides to stop until it actually does stop, including the reaction time of the driver (thinking distance) and the braking distance. *If you double your speed then the kinetic energy of your car increases by four (k.e. $= \frac{1}{2}mv^2$). This means the work done by the brakes also increases fourfold, so the distance you travel with the brakes applied before you stop is four times greater.*

terminal velocity is the constant velocity reached by an object falling through a fluid (liquid or gas) when its gravitational force (weight) is equal to the frictional forces acting on it. *Overall the net resultant force on an object travelling at terminal velocity is zero. All free-falling objects will reach a terminal velocity.*

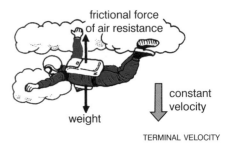

frictional force of air resistance

weight

constant velocity

TERMINAL VELOCITY

The size of this terminal velocity depends on their shape and area.

ticker-timer A ticker-timer is a device which is used to study motion. *The moving object (trolley) is attached to a paper tape which is pulled through a timer that prints a dot on the tape every $\frac{1}{50}$th of a second. The distance between the dots on the tape indicates the speed at which the object is moving. If the dots are equally spaced then the object is moving at constant speed. If the distance between the dots increases the object is accelerating; if the distance decreases the object is decelerating.*

distance–time graph A distance–time graph is one showing the change in distance with time. *The speed at a particular time is equal to the gradient (slope) of the graph at that point.*

acceleration is the rate of change of increasing velocity (or speed). *If the velocity changes at a constant rate then the object travels at constant or uniform acceleration. Acceleration, like velocity, is a **vector quantity** as it has both size and direction. Acceleration is often expressed in units of metres per second per second ($m\,s^{-2}$).*

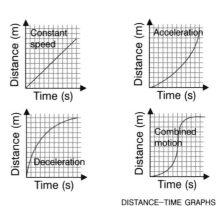

DISTANCE–TIME GRAPHS

deceleration is the rate of change of decreasing velocity (speed).

equations of motion are equations used to solve problems involving moving objects.

$$v = u + at$$
$$s = \tfrac{1}{2}(v + u)t$$
$$s = ut + \tfrac{1}{2}at^2$$
$$v^2 = u^2 + 2as$$

s = distance (m)
v = final velocity ($m\,s^{-1}$)
u = initial velocity ($m\,s^{-1}$)
a = acceleration ($m\,s^{-2}$)
t = time (s)

$$speed = \frac{distance\ travelled}{time\ taken}$$

$$acceleration = \frac{change\ in\ speed}{time\ taken}$$

velocity–time graph A velocity–time graph is one showing the change in velocity with time. *The acceleration at a particular time is equal to the gradient (slope) of the graph at that point. The area underneath a velocity–time graph is numerically equal to the total distance travelled.*

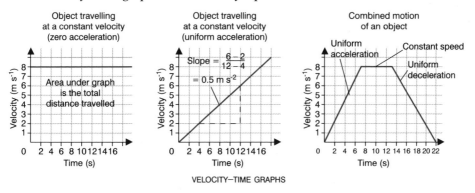

VELOCITY–TIME GRAPHS

force and acceleration are directly proportional and are related by the equation:

$$force = mass \times acceleration$$

When you double the force acting on an object you will double its acceleration (as long as its mass remains constant).

acceleration of gravity is the acceleration which the gravitational pull of the Earth exerts on a freely falling object. *On the Earth's surface it has a value of 9.8 $m\,s^{-2}$. This means that for every second, the object's velocity increases by 9.8 $m\,s^{-1}$.*

motion (periodic and circular)

periodic motion is any motion which is continuous and which repeats itself at regular intervals. *Examples are the motion of waves, the swinging action of a pendulum, and other **oscillations**.*

oscillation An oscillation is periodic motion between two extremes about a mean position, like a weight on the end of a spring oscillating up and down. *In oscillating systems there is a continuous change between **kinetic energy** and **potential energy**. The total sum of these energies remains constant (if there is no **damping**).*

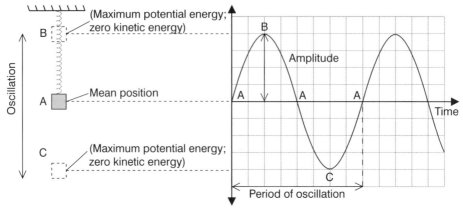

OSCILLATION OF WEIGHT ON A SPRING

mean position The mean position is the position about which an object will oscillate, and is the rest position of the object when the oscillation ceases. *Passing through this position during an oscillation, the object has maximum kinetic energy as it has its greatest velocity. At the same time, the object has its minimum potential energy as it has zero displacement.*

amplitude (symbol: *a*) The amplitude is the maximum displacement of an oscillating object from its mean position.

cycle One cycle is one complete motion. *For example, one complete oscillation, or one complete orbit, or one complete rotation of a spinning object.*

period (symbol: *T*) The period is the time taken to complete one cycle of motion. *For example, the period of rotation of the earth around the Sun is $365\frac{1}{4}$ days.*

frequency (symbol: *f*) The frequency is the number of complete cycles of a motion in one second. *The SI unit of frequency is hertz (Hz) which is equal to one cycle per second (c.p.s.). Frequency is also the reciprocal of the **period** of the motion.*

$$\text{frequency} = \frac{1}{\text{period}}$$
$$f = \frac{1}{T}$$

damping is the decrease in the amplitude of an oscillating system due to energy being drained away, e.g. as waste heat in overcoming friction or other resistive

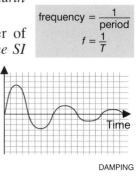

DAMPING

forces. *An example of damping is the action of the shock absorbers on a car. These allow the oscillations of the car, after going over a bump, to die down as quickly as possible.*

circular motion is motion of an object in a circle. *If the speed of the object remains the same then it is uniform circular motion. However the velocity does change, as the direction of the velocity is continually changing (though not its magnitude). This means the object is constantly accelerating towards the centre, so there is a force acting towards the centre. This is called the* **centripetal force.**

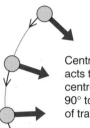

Centripetal force acts towards the centre and is at 90° to the direction of travel

CIRCULAR MOTION

centripetal force is the force directed toward the centre that causes a body to move in a uniform circular path. *This force depends on several factors. A larger centripetal force is needed for a ball to follow a circular path if*
- *the mass of the ball is increased*
- *the speed of the ball is increased*
- *the radius of the circle is decreased.*

These relationships are shown by the following equation:

$$\text{centripetal force} = \frac{\text{mass} \times (\text{velocity})^2}{\text{radius of circle}}$$

This inward centripetal force has no effect on the speed of the ball as it acts at right angles to the direction of motion.

satellites in orbit The gravitational pull (weight of satellite) provides the centripetal force needed to keep a satellite in a circular path around the Earth. *For a satellite to stay in a particular orbit it must travel at a certain speed. If it travels at less than this speed it will spiral inwards towards Earth. If it travels at greater than this speed it could escape the gravitational pull and go out into space. For a satellite orbiting just above the atmosphere the orbital speed is 8 km s^{-1}. The 'burn time' of the launch vehicle must be carefully controlled to ensure this speed.*

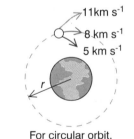

For circular orbit, gravitational acceleration must equal centripetal acceleration

$$\cancel{m}g = \frac{\cancel{m}v^2}{r}$$

$$g = \frac{v^2}{r}$$

SATELLITES IN ORBIT

escape velocity (or **escape speed**) is the minimum speed a satellite must travel to escape the Earth's gravitational pull. *For example, a rocket must travel at speeds of greater than 11.2 km s^{-1} to escape into space. This velocity is such that it gives a greater* **kinetic energy** *to the rocket than the* **potential energy** *resulting from the gravitational pull of the Earth.*

natural cycles

natural cycles are never-ending series of processes which maintain a balance in the environment of important substances that are essential for all plant and animal life. *Atoms cannot be destroyed during biological and chemical processes on Earth, and are therefore constantly recycled.*

water cycle (or **hydrological cycle**) The water cycle is the constant circulation of water between the atmosphere, the land, and the oceans. *Evaporation from oceans, lakes, and vegetation produces water vapour which forms clouds in the atmosphere. The water in the clouds then falls to the ground by precipitation (rain, snow, hail). Some percolates through the soil as groundwater to underground streams.*

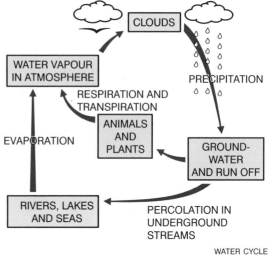

WATER CYCLE

*Some drains off the surface direct into streams and rivers. All this water eventually drains back into the seas and oceans for the cycle to begin again. Some water on the ground is taken in by plants and animals. This is returned to the atmosphere as water vapour by **respiration** and by **transpiration** in plants.*

carbon cycle The carbon cycle is the constant circulation of carbon between the atmosphere, plants, animals, and the soil. *Carbon dioxide is taken from the atmosphere during **photosynthesis** and incorporated into plant tissue. Animals feeding on plants then incorporate the carbon into their tissue. **Respiration** by both plants and animals returns carbon back into the atmosphere as carbon*

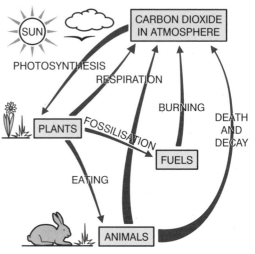

CARBON CYCLE

dioxide. Dead organic matter (plant and animal remains) undergoes bacterial decay, with respiration of the decomposers also releasing carbon

160

dioxide. *Some tissue in plants becomes fossilised into **fossil fuels**. **Combustion** of such fossil fuels releases carbon dioxide back into the atmosphere. The excessive use of fossil fuels contributes to **global warming**, as carbon dioxide is a **greenhouse gas**.*

nitrogen cycle The nitrogen cycle is the constant circulation of nitrogen between the atmosphere, plants, animals, and the soil. *Atmospheric **nitrogen gas** is inert and insoluble and cannot be used directly by plants. However, some of this gas undergoes **nitrogen fixation** by lightning or the action of nitrogen-fixing bacteria in the soil, which converts nitrogen into nitrates that dissolve in rain water. **Nitrates** in the soil are taken up by plants through their roots to make plant **proteins**. Animals feeding on the plants convert plant proteins into animal proteins. After death, the nitrogen-containing proteins of plants and animals are broken down by bacterial decay into **ammonia**, which can then be converted by other nitrifying bacteria back into nitrates. Some of the nitrogen in the nitrates is returned directly back into the atmosphere as nitrogen gas by denitrifying bacteria. Artificial **fertilisers** also become taken up in the nitrogen cycle, but may also contribute to **nitrate pollution**.*

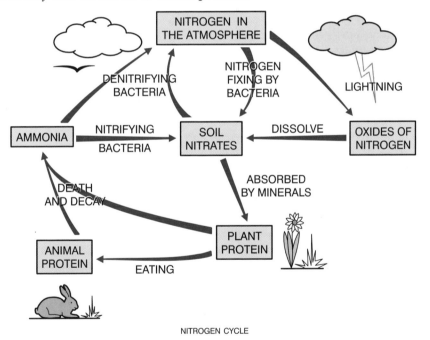

NITROGEN CYCLE

nitrogen fixation is the chemical process by which atmospheric **nitrogen** gas is converted into nitrogen compounds and enters the **nitrogen cycle**. *It is performed by bacteria in the soil, and in nodules on the root of leguminous plants (e.g. peas and beans) which contain symbiotic bacteria. Nitrogen fixation also occurs when the electrical discharge of lightning combines the nitrogen and oxygen in the atmosphere.*

nervous system A nervous system is that part of an organism which allows it to detect its surroundings and to react accordingly.

central nervous system (abbreviation: **CNS**) The central nervous system is that part of the nervous system that coordinates and controls all of the neural activity in an organism. *In vertebrates it consists of the **brain** and the **spinal cord**. The **CNS** processes information from the sense organs and produces a response.*

brain The brain is the main organ in the central nervous system which coordinates and controls most nerve activity. *It is protected by the **cranium** and is surrounded by three linings called **meninges**. It is made up of millions of nerve cells which are arranged in various sensory and motor areas. The main parts of the human brain are the highly developed **cerebrum**, and the **cerebellum**, **medulla oblongata**, and **hypothalamus**.*

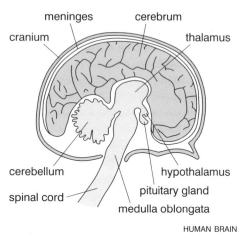

HUMAN BRAIN

cerebrum The cerebrum is the largest and most highly developed area of the forebrain and it controls most physical activity and all intelligent action such as speech, learning, decision making, imagination, etc. *It is composed of two **cerebral hemispheres**.*

cerebral hemispheres are the two halves of the cerebrum which form the main sensory and motor areas of the brain.

cerebellum The cerebellum is the front part of the hindbrain which coordinates and controls muscle movement and balance.

medulla oblongata The medulla oblongata is the posterior part of the brain which controls all reflex actions and involuntary actions like the heartbeat, breathing rate, blood pressure, etc. *It is under the overall control of the **hypothalamus**.*

hypothalamus The hypothalamus is the master controller of all unconscious activity and is situated at the base of the brain. *It controls the **autonomic nervous system** and the action of the **pituitary gland**. It also controls water and temperature regulation and so is important in **homeostasis**.*

pituitary gland (or **master gland**) The pituitary gland is at the base of the brain and controls the production of **hormones** by the **endocrine glands**.

spinal cord The spinal cord is that part of the **central nervous system** which runs down from the brain and is enclosed within the **vertebral column** (spine). *It consists of nervous tissue which is connected to **receptors** and* *effectors in the other parts of the body. The spinal cord is involved in **reflex** or **involuntary actions**.*

cerebrospinal fluid (abbreviation: **CSF**) is a fluid similar to **lymph** which fills the cavities in the brain and spinal cord, nourishing and protecting the tissues.

meninges (singular: **meninx**) are three protective membranes that cover the surface of the central nervous system in vertebrates. *Inflammation of these membranes is called meningitis.*

- **dura mater** is the tough outer **meninx** which protects the delicate inner meninges.
- **arachnoid membrane** is the middle **meninx** which carries **cerebrospinal fluid** and cushions the nervous tissue.
- **pia mater** is the innermost **meninx** which secretes cerebrospinal fluid.

peripheral nervous system (abbreviation: **PNS**) The peripheral nervous system is that part of the nervous system other than the central nervous system. *It includes all the nerves outside the brain and spinal cord.*

receptor A receptor is a sense organ like the eyes, ears, nose, skin, taste buds, etc. which can detect information about the surroundings. *Information is transmitted from receptors by **sensory neurones**.*

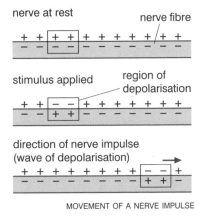

COMPONENTS OF THE NERVOUS SYSTEM

effector An effector is an organ such as a muscle or gland which can produce a response to a particular situation. *Information is transmitted to effectors by **motor neurones**.*

nerve impulse A nerve impulse is an electrical signal which moves along a **nerve fibre**. *The surface of a nerve fibre is positively charged. Stimulating the surface mechanically (as in the ear), electrically, or chemically causes the positive charge to be reversed (depolarisation). Movement of this depolarisation along the nerve fibre corresponds to the movement of the nervous (electrical) impulse.*

MOVEMENT OF A NERVE IMPULSE

neurone (or **nerve cell**) A neurone is an elongated branched cell that is the basic unit of the nervous system. *Neurones are specialised cells which can transmit electrical messages or nerve impulses around the body. There are three main types of neurone:* **sensory neurone**, **motor neurone**, *and* **connecting neurone**.

nerve fibre A nerve fibre is a bundle of **axons** or **dendrons** of nerve cells (neurones).

sensory neurones (or **afferent neurones**) transmit information as nervous impulses from **receptors** to the central nervous system. *Most sensory neurones have a long single* **dendron** *and a short* **axon** *(see diagram).*

dendron (or **dendrite**) A dendron is the part of the **neurone** that conducts electrical impulses towards the cell body of the nerve. *Branching dendrons are called dendrites.*

axon The axon is the part of the **neurone** that conducts electrical impulses away from the cell body of the nerve.

connecting neurones (or **relay neurones**) pick up information from sensory neurones, and pass new nervous impulses to the motor neurones to initiate a response. *Connecting neurones are often found in the* **central nervous system**.

motor neurones (or **efferent neurones**) transmit information as nervous impulses from the central nervous system to **effectors**. *Most motor neurones have branching* **dendrons** *(dendrites) and a long* **axon** *(see diagram) which is the nerve fibre.*

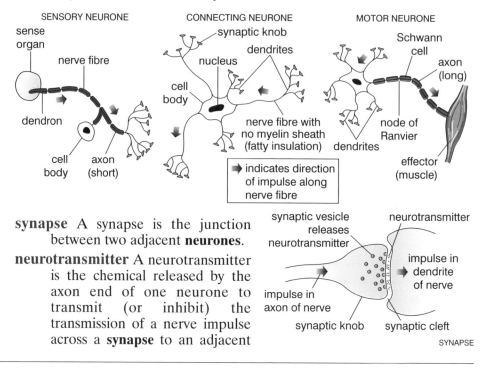

synapse A synapse is the junction between two adjacent **neurones**.

neurotransmitter A neurotransmitter is the chemical released by the axon end of one neurone to transmit (or inhibit) the transmission of a nerve impulse across a **synapse** to an adjacent neurone. *When such chemicals (like acetylcholine or noradrenalin) reach the dendron of the adjacent neurone the nerve impulse is triggered.*

voluntary actions are actions which are controlled by conscious activity of the brain. *Most voluntary actions involve **voluntary muscles** as **effectors**, often called skeletal muscles as they are attached to the skeleton. Information from the brain is carried to such muscles by **motor neurones**.*

involuntary actions are actions which are not controlled by conscious activity of the brain. *Such actions are controlled by the **hypothalamus** through neurones of the **autonomic nervous system**, and the **effectors** are **involuntary muscles**. They include gland secretion, heartbeat, peristalsis, accommodation of the eye, pupil contraction/dilation, etc.*

autonomic nervous system (abbreviation: **ANS**) The autonomic nervous system is that part of the nervous system that is not under voluntary control. *It is controlled by the **hypothalamus**, which sends information via **motor neurones** to **involuntary muscles** (smooth and cardiac muscles) and to the glands of the body.*

reflex actions are special types of involuntary action of which we are aware like swallowing, coughing, etc. *Such actions follow a 'neural short circuit' called a reflex arc.*

cranial reflexes are **reflex actions** of the head like sneezing and blinking, where the reflex arc goes only through the small section of the brain.

spinal reflexes are **reflex actions** where the reflex arc is through the spine. *Sensory information is fed through spinal nerves going into the dorsal (back) side of the spine. Motor nerves come from the ventral (front) side. Often spinal reflexes involve withdrawal away from a painful stimulus such as a hot surface (see diagram).*

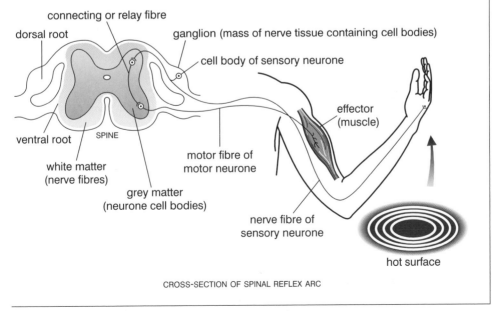

CROSS-SECTION OF SPINAL REFLEX ARC

carbon (chemical symbol: C) is the lightest non-metallic element in **group IV** of the periodic table and forms the basis of life chemistry. *It is an element which forms **allotropes** (e.g. graphite, diamond).*

carbon dioxide is the gaseous higher oxide of **carbon** which forms when carbon or its compounds burn in a plentiful supply of air (oxygen).

carbon monoxide is the gaseous lower oxide of **carbon** which forms when carbon or its compounds burn in a restricted supply of air (oxygen).

Carbon dioxide	Carbon monoxide
Odourless, colourless gas	Odourless, colourless gas
Turns lime water milky	No effect on limewater
Non-toxic	Very poisonous
Acidic oxide (dissolves to form carbonic acid)	Neutral oxide
Does not support combustion	Burns with a blue flame to form carbon dioxide
Not a reducing agent	Reducing agent

carbonates are ionic compounds containing the carbonate ion CO_3^{2-} (e.g. calcium carbonate $CaCO_3$). *All carbonates react with **acids** to form salts, water, and **carbon dioxide** gas. On heating most carbonates (except those of group I) undergo thermal **decomposition** to form the metal oxide and **carbon dioxide** gas. Most carbonates are insoluble in water (except group I).*

coal is a black hard mineral consisting mainly of carbon. *It is a **fossil fuel** and is a source of various organic chemicals.*

coke is the residue left behind after the destructive distillation of coal. *It is a greyish, brittle, porous solid containing about 85% carbon.*

charcoal is a porous form of carbon made when organic material is heated with very little air. *All forms of charcoal are porous, and are good at absorbing gases and purifying liquids (e.g. in sugar refining). Activated charcoal has been activated for absorption by heating in a vacuum. Uses of charcoal include as a smokeless fuel for barbecues, for drawing by artists, and for absorbing odours in shoe linings, gas masks, etc.*

wood charcoal is formed by heating wood in the absence of air. *It is used as a fuel.*

animal charcoal is made by heating bones and dissolving out the mineral salts with acid. *It is used in sugar refining.*

carbon black (or soot) is a fine powdered **amorphous** (non-crystalline) form of carbon formed by burning hydrocarbons in insufficient air. *It is used as a pigment and filler (e.g. for rubber).*

amorphous describes a solid which has no crystalline structure. *Examples include **carbon black** and **plastic sulphur**.*

carbon fibres are the black silky threads of carbon formed by charring textile fibres at temperatures from 700°C to 1800°C. *They are light and strong and are used to reinforce plastic resins to make high strength composites for fishing rods, squash and tennis rackets, golf clubs, etc.*

sulphur (chemical symbol: S) is a yellow non-metallic element found in group VI of the Periodic Table. *It is an element which forms **allotropes**.*

rhombic sulphur (or **alpha sulphur** or **orthorhombic sulphur**) is a pale yellow crystalline **allotrope** of sulphur which is stable at room temperature.

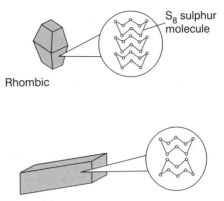

Rhombic

S_8 sulphur molecule

monoclinic sulphur (or **beta sulphur**) is a yellow crystalline **allotrope** of sulphur which is stable at temperatures above 96°C.

plastic sulphur is an **amorphous** form of sulphur formed when molten sulphur (m.p. 113°C) is poured into cold water. *It forms into long fibres which can be stretched and pulled like plastic. It is not stable and eventually crystallises to S_8 molecules.*

Monoclinic

CRYSTALLINE FORMS OF SULPHUR

flowers of sulphur is an **amorphous** form of sulphur formed when sulphur vapour (b.p. 445°C) is cooled quickly. *It is used as a plant fungicide.*

sulphur dioxide (or **sulphur (IV) oxide**) is a pungent smelling gas, SO_2, formed when sulphur burns in air. *It is a **reducing agent** and dissolves in water to form **sulphurous** acid. It is used in bleaching, as a fumigant, and in food preserving. Large quantities are used in the **contact process**.*

sulphur trioxide (or **sulphur (VI) oxide**) is a white crystalline solid, SO_3, which is formed in the **contact process**. *It is very volatile (m.p. 17°C) and reacts violently with water to form **sulphuric acid**.*

hydrogen sulphide is a colourless, poisonous gas, H_2S, which smells of 'bad eggs' and is made in 'stink bombs'. It is a weak acid and forms salts called **sulphides**.

sulphides are binary compounds (made of two elements) one of which is sulphur: e.g. iron(II) sulphide, FeS. *Many sulphides are ores.*

Acid	Anion in salts
hydrogen sulphide H_2S	sulphide S^{2-}
sulphurous acid H_2SO_3	sulphite SO_3^{2-}
sulphuric acid H_2SO_4	sulphate SO_4^{2-}

sulphites (or **sulphate (IV) salts**) are ionic compounds containing the sulphite ion SO_3^{2-} (e.g. sodium sulphite Na_2SO_3).

sulphates (or **sulphate (VI) salts**) are ionic compounds containing the sulphate ion SO_4^{2-} (e.g. sodium sulphate Na_2SO_4).

sulphurous acid (or **sulphuric (IV) acid**) is a colourless weak acid, H_2SO_3, formed when **sulphur dioxide** dissolves in water.

sulphuric acid (or **sulphuric (VI) acid**) is an oily, very corrosive **mineral acid**, H_2SO_4, formed when **sulphur trioxide** reacts with water. *When concentrated it is a dehydrating agent and **oxidising agent**. Commercially it is very important and is made on a large scale by the **contact process**.*

nitrogen (chemical symbol: **N**) is the first element in **group V** of the periodic table and is a **non-metal**. *It is a colourless, odourless gas (N_2) which makes up 78% of the **air**. It can be separated from the other gases in the air by **fractional distillation**. It is an unreactive gas but does have some uses, especially in the manufacture of **ammonia** by the **Haber process**.*

ammonia is a colourless, pungent gas, NH_3, that is less dense than air. *It is the most soluble of all gases and dissolves in water to form an **alkali** called aqueous ammonia, $NH_3(aq)$. It is the only common alkaline gas and moist red litmus paper immediately turns blue in the presence of ammonia. It neutralises acids to form **salts**. For example:*

> ammonia + sulphuric acid $\rightarrow$ ammonium sulphate
>
> $2NH_3$ + H_2SO_4 $\rightarrow$ $(NH_4)_2 SO_4$

*Commercially ammonia is very important, with million of tonnes being made each year by the **Haber process**, mostly for the manufacture of **fertiliser**.*

fertiliser A fertiliser is any substance that is added to the soil to increase the fertility of the soil.

natural fertilisers are of natural organic origin and consist of animal or plant remains. *Natural animal fertilisers are called manure and are rich in **trace elements**. Natural plant fertilisers are called compost.*

synthetic fertilisers are artificially manufactured fertilisers and are normally inorganic. *The most important are nitrogenous fertilisers, which are normally made by converting **ammonia** into compounds such as ammonium nitrate NH_4NO_3 or ammonium sulphate $(NH_4)_2SO_4$. These are solids, for ease of handling, and are water-soluble so they seep into the soil to be absorbed by the roots of the plant. The proportion of nitrogen present in a particular fertiliser is usually marked on the bag (see **NPK values**).*

NITROGENOUS FERTILISERS

Fertiliser	Formula	M_r	Proportion of nitrogen		
ammonium sulphate	$(NH_4)_2SO_4$	132	28/132	=	21.2%
ammonium nitrate	NH_4NO_3	80	28/80	=	35%
urea	NH_2CONH_2	60	28/60	=	46.6%

NPK values are numbers on a fertiliser bag which show the percentage by mass of the three important elements **nitrogen**, **phosphorus**, and **potassium** which are needed for healthy plant growth.

trace elements are elements which are only needed in small amounts but are still essential for healthy plant growth. *These include calcium, magnesium, sodium, and sulphur, as well as very small amounts of iron, copper, zinc, molybdenum, and cobalt.*

oxides of nitrogen $(NO)_x$ are compounds such as nitrogen monoxide NO and nitrogen dioxide NO_2 which are emitted from car exhausts, aircraft, and factories and cause **pollution**.

MAIN ELEMENTS NEEDED FOR HEALTHY PLANT GROWTH (apart from carbon, hydrogen and oxygen)

Element	Why plants need it	Signs of deficiency
Nitrogen	To make proteins which are needed for healthy growth of stems and leaves	Undersized leaves and slow growth
Phosphorus	For good root growth and to make DNA	Stunted growth causing purple young leaves
Potassium	Helps enzymes in photosynthesis and respiration	Yellow leaves with dead spots

nitric acid (or **nitric (V) acid**) is a colourless, extremely corrosive **mineral acid** HNO_3. *Industrially it is made by the Ostwald process. The main large-scale uses of nitric acid are in making fertilisers, explosives, and dyes.*

nitrous acid (or **nitric(III) acid**) is a colourless, weak acid HNO_2 which is easily decomposed. *Salts of the acid are called nitrites.*

Ostwald process The Ostwald process is the industrial synthesis of **nitric acid** by the catalytic oxidation of **ammonia**, and dissolving the resulting **oxides of nitrogen** in oxygenated water.

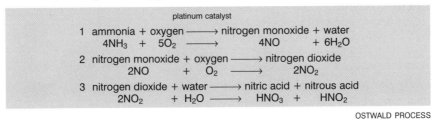

platinum catalyst

1 ammonia + oxygen ⟶ nitrogen monoxide + water
$$4NH_3 + 5O_2 \longrightarrow 4NO + 6H_2O$$

2 nitrogen monoxide + oxygen ⟶ nitrogen dioxide
$$2NO + O_2 \longrightarrow 2NO_2$$

3 nitrogen dioxide + water ⟶ nitric acid + nitrous acid
$$2NO_2 + H_2O \longrightarrow HNO_3 + HNO_2$$

OSTWALD PROCESS

nitrates (or **nitrate (V) salts**) are ionic compounds containing the nitrate ion NO_3^-: e.g. potassium nitrate KNO_3. *Nitrates are the salts of nitric acid. Ammonium nitrate is an important fertiliser. Sodium and potassium nitrates are used in explosives such as gunpowder.*

phosphorus (chemical symbol: **P**) is the second element in Group V of the Periodic Table and is a **non-metal**. It exists as several **allotropes**. *The main one is red phosphorus which is non-poisonous and not flammable. The other is white (or yellow) phosphorus which is poisonous, and is kept in water as it is spontaneously flammable in air. It is used in incendiary bombs and in the making of matches.*

phosphates (or **phosphate (V) salts**) are ionic compounds containing the phosphate ion PO_4^{3-}. *Phosphates are salts of phosphoric acid. Calcium phosphate is an important constituent of teeth and bones.*

phosphoric acid (or **phosphoric (V) acid**) is a tribasic acid H_3PO_4 which is a colourless crystalline solid. *It may be made by dissolving phosphoric (V) oxide in water. Phosphoric acid is used to form a corrosion-resistant layer on steel.*

phosphoric (V) oxide (or **phosphorus pentoxide**) is a white solid P_4O_{10} made by burning phosphorus in a plentiful supply of air. *It is a dehydrating agent.*

nuclear fission and fusion

nuclear energy is the energy released by **nuclear fission** or **nuclear fusion**. *When small atoms are fused together there is a small loss of mass which is released as energy (nuclear fusion). Atoms heavier than iron have added mass and will release energy if they break down (nuclear fission).*

nuclear reaction A nuclear reaction is any reaction in which there is a change in the nucleus of an atom.

mass defect is the difference in mass between the sum of individual masses of the protons and neutrons, and their total mass in the nucleus. *It is the mass equivalent of the binding energy of the mass–energy equation.*

binding energy is the energy required to cause a nucleus to decompose into its constituent neutrons and protons.

mass–energy equation shows the relationship $\Delta E = \Delta mc^2$, where ΔE = change in energy, Δm = change in mass and c is the speed of light $(3 \times 10^8\,\text{m}\,\text{s}^{-1})$. *A very small change in mass results in a vast change in energy (multiplication factor of around 10^{17}).*

nuclear fission is the process by which a heavy unstable nucleus is split up into two or more smaller nuclei called fission products. *This releases vast amounts of energy and emits two or three neutrons called fission neutrons. Most fission is induced by firing high energy neutrons at unstable nuclei (e.g. uranium-235 or plutonium-239). The neutrons released by induced fission will cause more fission and so on. This is a nuclear chain reaction.*

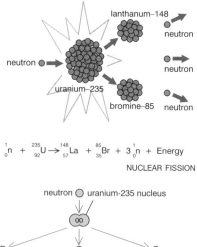

$$^{1}_{0}\text{n} + {}^{235}_{92}\text{U} \rightarrow {}^{148}_{57}\text{La} + {}^{85}_{35}\text{Br} + 3\,{}^{1}_{0}\text{n} + \text{Energy}$$

NUCLEAR FISSION

nuclear chain reaction A nuclear chain reaction is one that is self-sustaining. *Such reactions occur when a single neutron splits up one nucleus to produce (for example) three neutrons. These neutrons may split up three other nuclei which then produce nine neutrons, and so on. A controlled chain reaction is allowed to occur in a fission reactor. An uncontrolled chain reaction occurs in a fission bomb.*

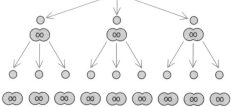

neutron ○ uranium-235 nucleus

NUCLEAR CHAIN REACTION

fissile material is material whose nuclei will undergo **nuclear fission**, either spontaneously or by induction through neutron bombardment. *The elements uranium-235 and plutonium-239 are fissile materials.*

critical mass The minimum mass of fissile material that will sustain a nuclear chain reaction. *If the mass of the fissile material is too small ('subcritical* *mass'), too many of the neutrons produced by the first fission escape from the surface into the atmosphere, so a chain reaction does not occur.*

fission reactor A fission reactor is a common type of nuclear reactor in which heat is produced by nuclear fission. *This heat produces steam, which drives a turbine in a nuclear power station to generate electricity.*

fission bomb (or **atomic bomb** or **A-bomb**) A fission bomb is one in which two or more subcritical masses of **fissile material** are brought together to make a mass in excess of the critical mass. *This results in an uncontrolled **nuclear chain reaction**, releasing huge amounts of energy.*

nuclear fusion is the collision and joining together of two light nuclei to form a heavier, more stable nucleus. *Nuclear fusion will only occur at extremely high temperatures (millions of degrees). It therefore only occurs naturally in the Sun (or other stars). Here hydrogen **isotopes** are fused together to form helium atoms with the release of vast amounts of energy. Such nuclear reactions requiring extremely high temperatures are called thermonuclear reactions.*

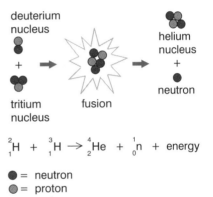

$$^{2}_{1}H + ^{3}_{1}H \rightarrow ^{4}_{2}He + ^{1}_{0}n + energy$$

● = neutron
○ = proton

NUCLEAR FUSION

fusion bomb (or **hydrogen bomb** or **H-bomb**) A fusion bomb is one in which a mixture of deuterium and tritium (isotopes of **hydrogen**) are fused together. *A 'trigger fission bomb' creates the very high temperatures needed for the fusion to take place. The energy released by this uncontrolled nuclear fusion is about 30 times greater than a conventional fission bomb of the same size.*

nuclear waste (or **radioactive waste**) is the waste from nuclear reactors, processing radioactive ores, reprocessing of nuclear fuel, and the manufacture of nuclear weapons. *Low-level waste (contaminated clothing, paper etc) may be buried about 100 m underground in stable geological formations. High-level radioactive waste (spent fuel rods, decommissioned reactor components, etc.) is extremely hazardous and can release radiation for thousands of years. It is vitrified in glass and then packed in stainless steel containers. These containers are then buried in thick concrete around 500 m underground in suitable dense rock.*

fall-out is radioactive particles that fall to Earth from the atmosphere. *Increased levels of radioactive fall-out result from a nuclear explosion. An accident in 1986 at the nuclear reactor at Chernobyl in the Ukraine resulted in fall-out over large populated areas. The most hazardous radioactive isotopes in fallout are strontium-90 and iodine-131. Both can be taken in by grazing animals (sheep, cows) and then passed on to humans in milk, milk products, and meat. Strontium-90 accumulates in the bones and iodine-131 in the thyroid gland.*

petroleum (or **crude oil** or **mineral oil**) is a naturally occurring mixture of organic compounds, mainly **hydrocarbons**, formed underground. *Microscopic organisms, which swam around in the oceans and seas around 300 million years ago, died and were buried in silt and mud. The pressure compressed the organic matter in their bodies, which gradually changed into petroleum. This brown/black liquid moved from the source rock to become trapped beneath layers of impermeable (non-porous) rock, floating on a layer of water and held under pressure below a layer of natural gas.*

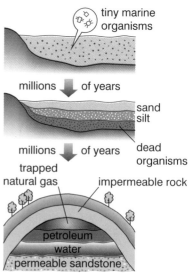

tiny marine organisms

millions of years

sand
silt

millions of years dead organisms

trapped natural gas

impermeable rock

petroleum
water

permeable sandstone

PETROLEUM FORMATION

natural gas is the gas that collects above underground deposits of oil. *It is a mixture of **hydrocarbons**, mainly methane.*

refining is the process of converting petroleum into more useful products. *Refining typically involves **fractional distillation** followed by the chemical process of **cracking**.*

fractional distillation (or **fractionation**) is a physical process used to separate a **mixture** of miscible liquids, such as petroleum, by distillation using a **fractionating column** (see diagram). *The petroleum is heated at the bottom of the column to around 400°C. It vaporises and is split into **fractions**. These travel up the column and become cooler. When a fraction reaches a tray at a temperature just below its boiling point, it condenses on to the tray and is drawn off along pipes.*

fraction A fraction is a mixture of liquids of similar boiling point. *Light fractions have lower boiling points and condense higher up the fractionating column than heavy fractions. Light fractions of petroleum are made of hydrocarbon molecules with small chains, while heavy fractions contain long hydrocarbon chains. Light fractions are volatile, flammable, pale-coloured liquids or gases, whereas heavy fractions are viscous (do not flow easily), less flammable, darker liquids. 90% of petroleum is used as a fuel and 10% as a chemical feedstock (similar to our food which is 90% fuel, 10% material for growth). Most fuels and feedstocks are the light fractions.*

petrochemicals are any of a range of various chemicals derived from petroleum or natural gas.

chemical feedstock describes fractions of petroleum which are used in the production of various organic chemicals. *Naphtha is the chief feedstock as it provides hundreds of very useful chemicals for paints, cosmetics, drugs, detergents, fuel additives, glues, pesticides, etc.*

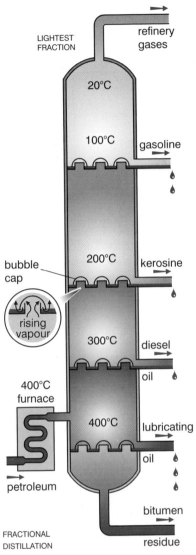

LIGHTEST
FRACTION

refinery
gases

20°C

100°C   gasoline

bubble
cap

200°C   kerosine

rising
vapour

300°C   diesel

oil

400°C
furnace

400°C   lubricating

oil

petroleum

bitumen

FRACTIONAL
DISTILLATION   residue

refinery gases (or **fuel gas**) is the petroleum fraction with a boiling point below 40°C. It is made up of a mixture of hydrocarbons with up to four carbon atoms (e.g. methane, ethane, propane, and butane). It is used as a fuel and kept liquefied in bottles as **LPG** (liquefied petroleum gas). This fraction is a chemical feedstock for making other organic chemicals.

gasoline (or **petrol** and **naphtha**) is the petroleum fraction with a boiling point range between 40–180°C. It is made up of a mixture of hydrocarbons containing between five and ten carbons. **Petrol** is a mixture of the hydrocarbons heptane C_7H_{14} and octane C_8H_{18} and is an important motor fuel. **Naphtha**, another mixture of hydrocarbons in this fraction, is the main **chemical feedstock** for making a wide range of chemicals including drugs, paint, and plastics.

kerosine (or **paraffin oil**) is the petroleum fraction with a boiling point range between 160–250°C. It is made up of a mixture of hydrocarbons which have 11 or 12 carbon atoms. Kerosine is a fuel for jet aircraft and oil-fired domestic heating. This fraction is also used for **cracking** to produce motor fuel.

diesel oil (or **gas oil**) is the petroleum fraction with a boiling point range between 220–350°C. It is made up of a mixture of hydrocarbons containing between 13 and 25 carbon atoms. Diesel oil is a fuel for diesel engines in lorries, ships, etc. It is called **DERV** (Diesel-Engine-Road Vehicles). This fraction can be 'cracked' to produce petrol.

lubricating oil is the petroleum fraction with a boiling point range 300–400°C. It is made up of a mixture of hydrocarbons containing 20 to 70 carbon atoms. It is used for lubricants, waxes, greases (Vaseline), and polishes. However, most of the fraction is used for **cracking**.

bitumen (or **asphalt**) is a semi-solid tarry substance left behind after distillation. It is commonly used in surfacing roads and in waterproofing felt for roofing.

cracking is the chemical process of breaking down large molecules from heavy fractions into more useful smaller molecules. *For example, the alkane hydrocarbons in the diesel oil fraction can be split into more useful hydrocarbons for petrol. Cracking also produces **unsaturated molecules** like ethene which are useful in plastic manufacture.*

| large alkane | → small alkane (petrol) | + alkene (butene) | + alkene (ethene) |
| $C_{16}H_{34}$ | C_8H_{18} | C_4H_8 | $2C_2H_4$ |

thermal cracking is the use of heat alone to break up heavy fractions.

catalytic cracking (or **cat cracking**) is the use of a catalyst together with heat to break up heavy fractions.

organic chemistry (hydrocarbons)

organic chemistry is the branch of chemistry concerned with the compounds of carbon (except carbonates and oxides of carbon). *'Organic' relates to living 'organisms', and all organic compounds are or have been associated with living material.*

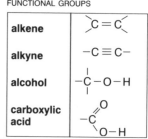

FUNCTIONAL GROUPS

hydrocarbons are organic compounds that contain only carbon and hydrogen atoms. *There are two main classes, **alkanes** and **alkenes**. All hydrocarbons have **covalent** molecules. They are found naturally in **petroleum** and **natural gas**.*

functional group A functional group is an atom or group of atoms that give an organic molecule its typical chemical properties. *Only alkanes do not have a functional group.*

homologous series A homologous series is a series of related organic compounds with the same **functional group**. *Members of such a series can be represented by a general formula. Each member of the series differs from the next by an additional $-CH_2-$ group. Members of a homologous series have similar chemical properties. Their physical properties gradually change as the molecule gets larger.*

alkanes (or **paraffin hydrocarbons**) are a homologous series of hydrocarbons with a general formula C_nH_{2n+2}. *Alkanes are **saturated molecules** which only contain single covalent bonds. They are the main hydrocarbons found in **petroleum** and **natural gas**.*

ALKYL GROUPS

alkyl group An alkyl group is a hydrocarbon radical (group of atoms) derived from an alkane by the removal of one hydrogen atom.

CH_3-	methyl
C_2H_5	ethyl
C_3H_7-	propyl
C_4H_9-	butyl

alkenes (or **olefin hydrocarbons**) are a homologous series of hydrocarbons with a general formula C_nH_{2n}. *Alkenes are **unsaturated molecules** with at least one carbon–carbon double bond. They are formed when petroleum fractions undergo **cracking**.*

n	Alkanes C_nH_{2n+2}			Alkenes C_nH_{2n}			Alkynes C_nH_{2n-2}		
	Name	Molecular formula	Structural formula	Name	Molecular formula	Structural formula	Name	Molecular formula	Structural formula
1	methane	CH_4							
2	ethane	C_2H_6		ethene	C_2H_4		ethyne	C_2H_2	$H-C\equiv C-H$
3	propane	C_3H_8		propene	C_3H_6		propyne	C_3H_4	$CH_3-C\equiv C-H$
4	butane	C_4H_{10}		butene	C_4H_4		butyne	C_4H_6	$C_2H_5-C\equiv C-H$
5	pentane	C_5H_{12}		pentene	C_5H_{10}		pentyne	C_5H_8	$C_3H_7-C\equiv C-H$
6	hexane	C_6H_{14}		hexene	C_6H_{12}		hexyne	C_6H_{10}	$C_4H_9-C\equiv C-H$

isomerism (or **structural isomerism**) occurs when compounds with the same molecular formula have different structural formula. *For example, butane and isobutane (methyl propane) are isomers.*

butane (C_4H_{10})

isobutane (C_4H_{10})

isomers are substances that exhibit isomerism. *Isomers have different physical properties but similar chemical properties.*

saturated molecules contain only **single covalent bonds** (e.g. **alkanes**).

unsaturated molecules contain **double covalent bonds** (alkenes) or **triple covalent bonds** (alkynes).

properties of alkanes Alkanes are fairly unreactive molecules as their C—C and C—H bonds are strong. *The first four members are gases. Higher members are liquids and eventually waxy solids. Alkanes burn in a plentiful supply of air to form carbon dioxide and water (steam).*

ethane	+	oxygen	$\rightarrow$	carbon dioxide	+	water
$2C_2H_6$	+	$7O_2$	$\rightarrow$	$4CO_2$	+	$6H_2O$

substitution reaction (or **replacement reaction**) A substitution reaction is a chemical reaction in which an atom or molecule is replaced by another atom or molecule. *Saturated molecules like alkanes undergo substitution reactions.*

methane	+	chlorine	$\rightarrow$	chloromethane	+	hydrogen chloride
CH_4	+	Cl_2	$\rightarrow$	CH_3Cl	+	HCl

properties of alkenes Alkenes are chemically more reactive than alkanes because of the carbon–carbon double bond. *Like alkanes, they burn in a plentiful supply of air to form carbon dioxide and water (steam).*

ethane	+	oxygen	$\rightarrow$	carbon dioxide	+	water
C_2H_4	+	$3O_2$	$\rightarrow$	$2CO_2$	+	$2H_2O$

addition reaction An addition reaction is a chemical reaction in which one molecule adds onto another. *Unsaturated molecules like alkenes undergo addition reactions.*

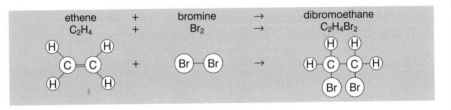

ethene	+	bromine	$\rightarrow$	dibromoethane
C_2H_4	+	Br_2	$\rightarrow$	$C_2H_4Br_2$

organic chemistry (alcohols)

alcohols are organic compounds that contains the **–OH functional group**. *They are usually colourless, flammable liquids which are good solvents and fuels. The most important is **ethanol**. Alcohols form a **homologous series** with a general formula* $C_nH_{2n+1}OH$.

HOMOLOGOUS SERIES OF ALCOHOLS

Value of n	Name of alcohol	Molecular formula	Structural formula	Boiling point/°C	Uses
1	Methanol	CH_3OH	H H–C–O–H H	66	1 Making meths 2 Solvent 3 Fuel
2	Ethanol	C_2H_5OH	H H H–C–C–O–H H H	78	1 Alcoholic drinks 2 Solvent/cosmetics 3 Fuel
3	Propanol	C_3H_7OH	H H H H–C–C–C–O–H H H H	97	1 Solvent 2 Aerosols 3 Antifreeze
4	Butanol	C_4H_9OH	H H H H H–C–C–C–C–O–H H H H H	118	1 Solvent 2 Perfumes (esters) 3 Flavouring (esters)

fermentation is the conversion of sugars into **ethanol** and carbon dioxide gas by the action of microorganisms such as **yeast**, in the absence of air. *This is the basis of baking, wine-making, and brewing beer. Fermentation is a type of **anaerobic respiration**, carried out by **enzymes** (zymase) in the yeast which work best at temperatures of around 25–30°C.*

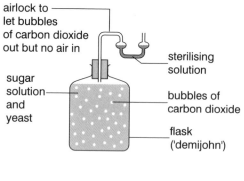

airlock to let bubbles of carbon dioxide out but no air in

sterilising solution

sugar solution and yeast

bubbles of carbon dioxide

flask ('demijohn')

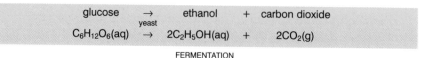

glucose $\rightarrow$ ethanol + carbon dioxide

$C_6H_{12}O_6(aq) \xrightarrow{yeast} 2C_2H_5OH(aq) + 2CO_2(g)$

FERMENTATION

ethanol (CH_3CH_2OH) is the commonest alcohol, and is a colourless, water-soluble liquid (boiling point 78°C). *It is formed in beers, wines, and spirits by the process of **fermentation**, which is also one method of large-scale production, if sugar is plentiful. However, it is sometimes more economical to produce ethanol from ethene gas obtained by cracking petroleum fractions. The reaction involves passing ethene and steam over a catalyst of phosphoric(V) acid at 300°C and 60 atmospheres pressure. The most important chemical properties of ethanol are as follows:*

C_2H_4 + H_2O → C_2H_5OH

ethene + water (steam) → ethanol

- *combustion* Like all alcohols ethanol burns in a plentiful supply of air to form carbon dioxide and steam. The reaction gives out lots of heat energy and is **exothermic.** Ethanol is sometimes used as a fuel in cars or rockets.

$$C_2H_5OH + 3O_2 \rightarrow 2CO_2 + 3H_2O$$
$$\text{ethanol} + \text{oxygen} \rightarrow \text{carbon} + \text{water}$$
$$\text{dioixde} \quad \text{(steam)}$$

- *oxidation* If left exposed in the air ethanol turns 'sour'. This is the common fate of wines and beers which are opened but not drunk. The reason is that the ethanol has been oxidised to **ethanoic acid** *(vinegar).*

$$C_2H_5OH + O_2 \rightarrow CH_3COOH + H_2O$$
$$\text{ethanol} + \text{oxygen} \rightarrow \text{ethanoic} + \text{water}$$
$$\text{acid}$$

methanol (CH_3OH) is the simplest alcohol, and is a good solvent and fuel, but is also poisonous.

methylated spirits (or **meths**) is a mixture of **ethanol** (90%) and **methanol** (9%) with small amounts of an emetic (to prevent it being drunk) and blue dye. *It is a common fuel and solvent.*

carboxylic acids (or **fatty acids**) are **organic acids** that have a —COOH functional group. *They are generally weak acids. This is because they exist mainly as molecules and do not form hydrogen ions as easily as mineral acids. However they do exhibit normal acidic properties.*

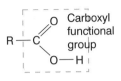

methanoic acid (or **formic acid**) is the simplest **carboxylic acid** HCOOH and occurs naturally in ants and stinging nettles. *It is formed by the oxidation of **methanol.***

ethanoic acid (or **acetic acid**) is a **carboxylic acid** formed by the oxidation of ethanol. *Vinegar contains 5% ethanoic acid. It is a weak **organic acid** and it is used for flavourings and as a preservative.*

esters are organic compounds formed by the reaction of a **carboxylic acid** and **alcohol.** *For example, if ethanoic acid is warmed with ethanol, in the presence of a few drops of concentrated sulphuric acid (hydrogen ions are the catalyst) then the ester ethyl ethanoate is formed. Esters are volatile fragrant substances used in perfumes and as flavourings in the food industry.*

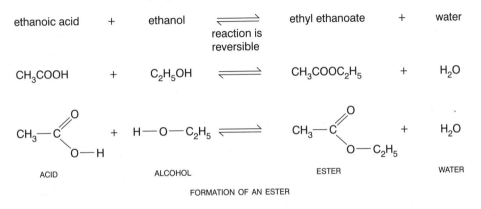

FORMATION OF AN ESTER

particles (states of matter)

kinetic theory states that matter is made up of particles which move with a vigour proportional to their absolute temperature. *In solids, particles vibrate about fixed positions; in liquids, they still vibrate but can move past each other; in gases, the particles are separate and able to move freely.*

states of matter are the three common physical forms or phases in which matter exists: **solid, liquid, gas**. *Substances can change between these states, usually when heated or cooled, as the **kinetic energy** (indicated by the velocity of the particles) increases or decreases respectively.*

KINETIC THEORY OF MATTER

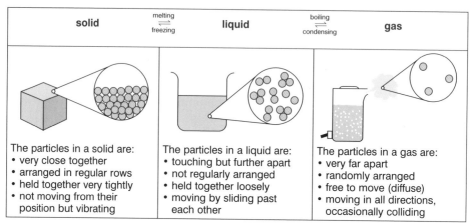

The particles in a solid are:
• very close together
• arranged in regular rows
• held together very tightly
• not moving from their position but vibrating

The particles in a liquid are:
• touching but further apart
• not regularly arranged
• held together loosely
• moving by sliding past each other

The particles in a gas are:
• very far apart
• randomly arranged
• free to move (diffuse)
• moving in all directions, occasionally colliding

plasma is a fourth **state of matter** which can only exist at very high temperatures, e.g. inside the Sun. *At such temperatures, matter is broken down into positive ions and electrons.*

fluid A fluid is a **state of matter** that can flow, like **gases** or **liquids**.

melting is the change in state from a solid to a liquid, usually caused by heating. *On heating, the particles in a solid gain energy and move more. They remain in contact but with weaker forces of attraction, and so they are not held in fixed positions and the material becomes a liquid.*

melting point is the temperature at which a **solid** completely changes into a **liquid**. *A pure substance has an exact melting point. Impurities cause lower melting points.*

molten describes the liquid state of a substance that is normally a solid at room temperature and pressure.

freezing is the solidification which occurs when a liquid is cooled. *The forces between the particles become stronger until they just vibrate about fixed positions and the material becomes a solid.*

freezing point is the temperature at which all of a liquid changes into a solid. *The freezing point of a pure substance is the same temperature as its melting point. Impurities lower (depress) freezing points.*

fusion is the change in state from a solid to liquid of a substance which is a solid at room temperature and pressure.

boiling is the rapid change in state from a liquid to a gas (or vapour) usually caused by heating. *On heating, the particles in a liquid gain energy. Eventually they have sufficient energy to overcome the binding forces in the liquid, and spread away as particles of a gas.*

boiling point is the temperature at which all of a **liquid** changes into a **gas** (or vapour) because the **vapour pressure** of the liquid is equal to atmospheric pressure. *A pure substance has an exact boiling point. Impurities raise (elevate) boiling points.*

evaporation is the process of a liquid changing into a vapour at temperatures below its boiling point. *It only occurs on the surface of a liquid when a particle has by chance sufficient energy to escape.*

vapour A vapour is a gas which is below its **critical temperature** and can be liquefied by pressure alone.

vapour pressure is the pressure of the vapour which is suspended above the surface of a liquid (or solid) at a particular temperature. *The liquid and the vapour are in a **phase equilibrium** with each other.*

volatile liquid A volatile liquid is one which has a low boiling point and changes easily into a vapour (high vapour pressure).

condensation is the change of state from gas (or vapour) to a liquid. *It is normally caused by cooling.*

sublimation is the direct change of state from a solid to a gas (or vapour) on heating or from a gas to a solid on cooling. *Iodine, solid carbon dioxide (dry ice) and some ammonium salts undergo sublimation.*

heating curve A heating curve is a graph showing changes in temperature with time for a substance being heated. *The flat sections on the graph indicate the melting and boiling points. Here the temperature remains the same over a period of time, as the heat energy is being used to change the structure of the substance.*

HEATING CURVE

cooling curve A cooling curve is a graph showing changes in temperature with time for a substance being cooled. *The shape is the same as a heating curve but in the opposite direction.*

Brownian motion is the random motion of particles in water or air caused by collision with the surrounding molecules. *It was first observed by Robert Brown in 1827 when studying pollen grains on the surface of water.*

diffusion is the mixing of two fluids without mechanical help. *Gases diffuse much faster than liquids as their particles move for longer distances before collision. Only **miscible** liquids can diffuse.*

particles (gases)

kinetic theory of gases states that in the gaseous state molecules of a gas are far apart and in random motion. *They travel in straight lines until they collide with other molecules or with the walls of the containing vessel. There are various gas laws which describe the behaviour of these gases. However these laws assume that the gas behaves as an **ideal gas.***

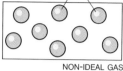

large particles
close together

NON-IDEAL GAS

ideal gas An ideal gas is a theoretical gas which obeys the various gas laws. *Many real gases behave in approximately the same way when at low/medium pressures and temperatures. The assumptions about an ideal gas are as follows:*

small particles
far apart

IDEAL GAS

- The size of the molecules in the gas are so small that their total volume is negligible.
- The molecules are so far apart that the attractive forces between the molecules is negligible.
- The molecules move in straight lines and lose no energy when they collide (elastic collisions).
- The molecules can convert heat energy into kinetic energy (and vice versa).

pressure of a gas (symbol: p) is the result of the continual collision of the molecules of the gas on the walls of the container. *It depends upon the temperature and volume of the gas.*

volume of a gas (symbol: V) is the amount of space the gas occupies, at a particular pressure and temperature.

temperature of a gas (symbol: T) is a measure of the average kinetic energy of the molecules. *It must be measured on the **absolute scale** in kelvins.*

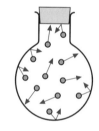

PRESSURE OF A GAS

Boyle's law states that the **volume** of a given mass of gas at a constant temperature is inversely proportional to its **pressure:** pV = constant. *This is only true for an **ideal gas**. The law was named after the Irish physicist Robert Boyle (1627–91). In terms of **kinetic theory**, when the volume of gas is halved, then there will be twice as many collisions on the walls of the container, if the mass remains the same. Therefore halving the volume of a gas doubles its pressure.*

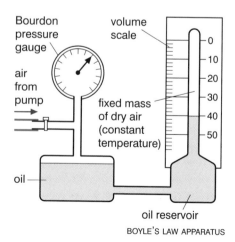

Bourdon pressure gauge

air from pump

oil

volume scale

fixed mass of dry air (constant temperature)

0
10
20
30
40
50

oil reservoir

BOYLE'S LAW APPARATUS

180

Charles' law states that the **volume** of a fixed mass of gas at constant pressure is directly proportional to its temperature (in kelvins): V/T = constant. *For an **ideal gas** the increase in volume for each degree rise in temperature is the fraction $\frac{1}{273}$ as long as the pressure of the gas remains unchanged. The law was named after the French scientist J.A.C. Charles (1746–1823).*

pressure law (or **Charles' law of pressure**) states that the **pressure** of a fixed mass of gas at constant volume is directly proportional to its temperature (in kelvins): p/T = constant. *This law only holds true with an **ideal gas** at low/medium pressures and temperatures. In terms of **kinetic theory**: when the temperature of the gas is increased the molecules travel faster with greater kinetic energy; they therefore collide with greater force on the walls of the container, so there is a greater pressure.*

Bourdon pressure gauge is an instrument for measuring the pressure of a gas or liquid. *It consists of a coiled flattened tube which straightens out as the pressure increases. The movement of the coiled tube is made to move a pointer across a scale to indicate the size of the pressure.*

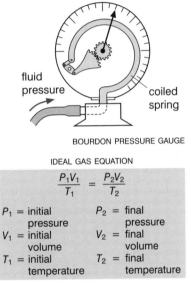

fluid pressure

coiled spring

BOURDON PRESSURE GAUGE

ideal gas equation is the combined equation of the three gas laws, which states that for a fixed mass of **ideal gas**, pV/T = constant.

IDEAL GAS EQUATION

$$\frac{P_1V_1}{T_1} = \frac{P_2V_2}{T_2}$$

P_1 = initial pressure P_2 = final pressure
V_1 = initial volume V_2 = final volume
T_1 = initial temperature T_2 = final temperature

critical temperature is the temperature above which a gas cannot be liquefied, no matter how high the pressure.

critical pressure is the pressure necessary to condense a gas at its critical temperature.

partial pressure is the pressure that each gas in a mixture of gases would exert if it alone filled the volume occupied by the mixture.

Dalton's law states that the total pressure exerted by a gaseous mixture is equal to the sum of the partial pressure of the gases. *Atmospheric pressure is 1 atmosphere, and since air is approximately 80% nitrogen and 20% oxygen, the partial pressures of nitrogen and oxygen are approximately 0.8 and 0.2 atmospheres respectively.*

standard temperature and pressure (abbreviation: **s.t.p.**) is the temperature of 0°C (273 K) and a pressure of 1 atmosphere (101 325 pascals). *One **mole** of any gas occupies a volume of 22.4 dm³ at standard temperature and pressure.*

room temperature and pressure (abbreviation: **r.t.p.**) is the temperature of 25°C (298 K) and a pressure of 1 atmosphere (101 325 pascals). *One **mole** of any gas occupies a volume of 24.0 dm³ at room temperature and pressure.*

periodic table of elements

periodic table The periodic table is an arrangement of **elements** in order of increasing number of protons (**atomic number**). *Periods and groups place together elements with related electronic configurations. The original form of the modern periodic table was first proposed by Dimitri Mendeleev in 1869.*

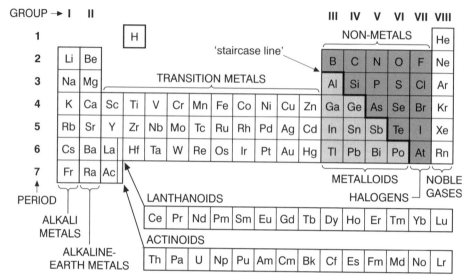

PERIODIC TABLE OF ELEMENTS

group A group is a vertical column of elements in the **periodic table**. *Groups are numbered using roman numerals. Elements within a particular group have similar chemical properties because they have the same number of **valence electrons**

Group number	Common name
I	Alkali metals
II	Alkaline-earth metals
VII	Halogens
VIII	Noble gases

(electrons in the outermost shell). Within groups of metallic element, the reactivity of the metal increases as we go down the group. Within groups of non-metallic elements, the reactivity of the non-metal decreases as we go down the group. Group IV contains **carbon** (the element of life) and **silicon** (the element of rocks).*

periodicity is the gradual change of physical properties (melting point, boiling point, etc.) for elements as we go across the **periodic table**. *The graph shows such periodicity in the boiling points of the first 60 elements.*

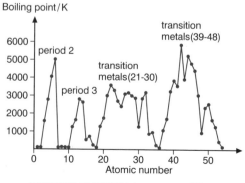

PERIODICITY OF BOILING POINT OF FIRST 50 ELEMENTS

period A period is a horizontal row of elements in the **periodic table**. *The first three rows are called 'short periods'. The next four rows, which include the **transition metals**, are called the 'long periods'. All elements within a period contain the same number of **electron shells**. As we go across a period there is a change from reactive **metals**, through less reactive metals, metalloids, and less reactive non-metals to reactive **non-metals**. On the extreme right are the **noble gases** (group VIII or group 0).*

metals are a class of chemical elements which always form positive ions (cations) when they react to form compounds. *They are often lustrous (shiny) solids which are good conductors of heat and electricity. They also form solid oxides that act as **bases**. In the Periodic Table, metallic elements are found on the left-hand side below the 'staircase-line'. As we go down a group the metallic character increases. As we go across a period the metallic character decreases. Caesium is the most reactive metallic element.*

transition metals are the block of metallic elements in the middle of the periodic table. *They have partly filled inner electron shells which gives them distinctive properties. They are typical metals. They are strong and hard, good conductors of heat and electricity, and have high melting points. Many of them have **variable valency** and coloured compounds, and act as **catalysts**.*

lanthanoids (or **rare-earth metals** or **lanthanide metals**) are a series of metallic elements of atomic numbers 57 to 71 inclusive. *The properties of these elements are very similar to the metal aluminium.*

actinoids (or **actinide metals**) are a series of metallic elements of atomic numbers 90 to 103 inclusive. *All are very rare metals and most are radioactive. Those metals after uranium are called 'transuranic elements' and are made in nuclear reactors.*

metalloids (or **semimetals**) are a class of elements intermediate in properties between metals and non-metals, such as boron and silicon. *They are often electrical **semiconductors** whose physical properties resemble **metals** but whose chemical properties resemble **non-metals**.*

semiconductor A semiconductor is a conductor whose electrical resistance decreases as the temperature rises. Semiconductors include **metalloids** like silicon, germanium, and their compounds. **Transistors, diodes**, and other electronic components are made from semiconductors.

non-metals are elements which do not have the properties of a metal and always form negative ions (anions) when they react to form ionic compounds. *Non-metallic elements are often gases (nitrogen, oxygen, fluorine, chlorine, and noble gases) or low melting point solids (phosphorus, sulphur, iodine). They are poor electrical and thermal conductors. Less than a quarter of the elements in the periodic table are non-metallic elements, and they are found on the right-hand side above the 'staircase-line'. As we go down a group the non-metallic character decreases. As we go across a period the non-metallic character increases. Fluorine is the most reactive non-metallic element.*

alkali metals (or **group I elements**) are the elements in the first group in the periodic table, which all have a single **valence electron**. *They all react with water to form **alkalis** (soluble metal hydroxides), hence their name.*

physical properties of the alkali metals show trends as we go down the group (see table). *Francium is excluded as it is an unstable radioactive element. The melting points and boiling points are much lower than you would normally associate with metals. They have weak interatomic forces, so they are also relatively soft metals and can be cut with a knife. The freshly cut surface is silvery but this soon tarnishes. The first three elements in the group are less dense than water and will therefore float on water.*

PHYSICAL PROPERTIES OF ALKALI METALS

Element	Symbol	Melting point (°C)	Boiling point (°C)	Atomic radius (nm)	Density (g/cm^{-3})	Appearance
Lithium	Li	180	1330	0.15	0.53	silvery-white metal
Sodium	Na	98	892	0.19	0.97	soft silvery-white metal
Potassium	K	64	760	0.23	0.86	soft silvery-white metal
Rubidium	Rb	39	688	0.25	1.53	soft silvery-white metal
Caesium	Cs	29	690	0.26	1.9	soft golden-coloured metal

chemical properties of the alkali metals are very similar and they are all very reactive, being the highest metals in the **reactivity series**. *This is because they have only one **valence electron** to lose to form the stable ion (Li^+, Na^+, K^+). Alkali metals are therefore good **reducing agents**. Because the elements are so reactive with oxygen and water, they are stored under oil to prevent reaction with the moist air. The metals all burn in air with coloured flames to form **basic oxides**. They all react violently with cold water to form **hydrogen** gas and an **alkali**. Alkali metals also react violently with **non-metals** like chlorine to form **ionic compounds**. Salts of alkali metals (nitrates, carbonates, sulphates, and chlorides) are white crystalline ionic solids which are water soluble.*

oil

metal

STORAGE OF ALKALI METALS

potassium burns with lilac flame

water   hydrogen gas

REACTION WITH WATER

occurrence Sodium and potassium salts are plentiful on the Earth's surface (sodium 2.7%, potassium 2.5% by mass in the Earth's crust).

Mineral	Main chemical constituent
common salt	NaCl (sodium chloride)
saltpetre	KNO_3 (potassium nitrate)
Chile saltpetre	$NaNO_3$ (sodium nitrate)

*Lithium salts are found only in trace amounts, and rubidium and caesium are extremely rare. As the salts of these metals are water-soluble there are high concentrations of these salts in the sea, especially **sodium chloride**.*

CHEMICAL PROPERTIES OF ALKALI METALS

Element	Symbol of ion	Flame colour	Chemical reaction with		
			oxygen	water	chlorine
Lithium	Li^+	crimson red	burns with red flame to form the oxide $4Li + O_2 \rightarrow 2Li_2O$	reacts with cold water to form the alkali and hydrogen gas $2Li + 2H_2O$ $\rightarrow 2LiOH + H_2$	burns to form the white chloride salt $2Li + Cl_2 \rightarrow 2LiCl$
Sodium	Na^+	brilliant yellow	burns with a yellow flame to form the oxide $4Na + O_2 \rightarrow 2Na_2O$	reacts violently to form the alkali and hydrogen gas $2Na + 2H_2O$ $\rightarrow 2NaOH + H_2$	burns to form the white chloride salt $2Na + Cl_2 \rightarrow 2NaCl$
Potassium	K^+	lilac	burns with a lilac flame to form the oxide $4K + O_2 \rightarrow 2K_2O$	catches fire and forms the alkali and hydrogen gas $2K + 2H_2O$ $\rightarrow 2KOH + H_2$	burns to form the white chloride salt $2K + Cl_2 \rightarrow 2KCl$

extraction of alkali metals is carried out by electrolysis of their molten chlorides. *The molten metal collects at the steel cathode, and chlorine gas collects at the graphite anode.*

> **cathode** $Na^+ + e^- \rightarrow Na$ (sodium metal)
> **anode** $2Cl^- \rightarrow Cl_2 + 2e^-$ (chlorine gas)

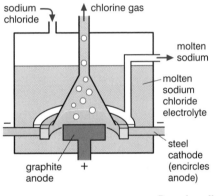

Down's cell

FOR EXTRACTION OF SODIUM BY ELECTROLYSIS

uses of sodium Sodium is the only alkali metal used on a commercial scale: the others are too expensive or too reactive. *Sodium is a much better thermal conductor than water and does not boil until 892°C. Molten sodium is therefore a very good liquid coolant (in the absence of air and water). It is used as such in certain types of nuclear reactor. Sodium vapour in street lamps gives them a characteristic yellow colour. Because sodium is so reactive, it can be used in the extraction of less reactive metals from their salts. One such **displacement reaction** is used to extract titanium metal:*

> titanium chloride + sodium → titanium + sodium chloride
> $TiCl_4$ + $4Na$ → Ti + $4NaCl$

Solvay process The Solvay process is used for the industrial manufacture of **sodium hydrogen carbonate** and **sodium carbonate** by bubbling ammonia and carbon dioxide gases through concentrated salt water.

> $NaCl + H_2O + NH_3 + CO_2 \rightarrow NH_4Cl + NaHCO_3$
> salt water   sodium hydrogen carbonate

periodic table (alkaline-earth metals)

alkaline-earth metals (or **group II elements**) are the elements in the second group in the periodic table, which all have two **valence electrons**. *They are called alkaline-earth because their oxides are alkaline, and because salts of calcium and magnesium are common in the Earth's crust.*

physical properties of the alkaline-earth metals are similar for all the elements (except beryllium). *The melting and boiling points are higher than the alkali metals (Group I) as there are two electrons per atom, so double the bonding strength. However, they are still lower than a typical transition metal. The alkaline-earth metals are noticeably softer than transition metals and have lower densities.*

PHYSICAL PROPERTIES OF ALKALINE-EARTH METALS

Element	Symbol	Melting point (°C)	Boiling point (°C)	Atomic radius (nm)	Density (g cm^{-3})	Appearance
Beryllium	Be	1280	2270	0.11	1.85	hard white metal
Magnesium	Mg	650	1110	0.16	1.74	quite soft silvery metal
Calcium	Ca	838	1440	0.2	1.55	soft silvery metal
Strontium	Sr	768	1380	0.21	2.6	soft silvery metal
Barium	Ba	714	1640	0.22	3.5	soft silvery metal

chemical properties of the alkaline-earth metals are very similar: they are all reactive metals whose reactivity increases down the group. *They always form the stable M^{2+} ion (Be^{2+}, Mg^{2+}, Ca^{2+}) because they lose two electrons to form the stable inert gas configuration. All burn in air with characteristic flame colours to form **basic oxides**. However, the alkaline hydroxides of group II are much less soluble than group I. Alkaline-earth metals react with **non-metals** like chlorine to form **ionic compounds**. Salts of alkaline-earth metals (nitrates, carbonates, sulphates, and chlorides) are white crystalline ionic solids. The solubility of the sulphates and carbonates decreases down the group.*

CHEMICAL PROPERTIES OF ALKALINE-EARTH METALS

Element	Symbol of ion	Flame colour	Chemical reaction with		
			oxygen	water	chlorine
Magnesium	Mg^{2+}	white	burns with brillliant white light to form oxide $2Mg + O_2 \rightarrow 2MgO$	reacts with steam to form oxide and hydrogen $Mg + H_2O(g)$ $\rightarrow MgO + H_2$	burns to form the white chloride salt $Mg + Cl_2 \rightarrow MgCl_2$
Calcium	Ca^{2+}	red	burns with red flame to form oxide $2Ca + O_2 \rightarrow 2CaO$	reacts with cold water to form hydroxide and hydrogen $Ca + 2H_2O(l)$ $\rightarrow Ca(OH)_2 + H_2$	burns to form the white chloride salt $Ca + Cl_2 \rightarrow CaCl_2$
Barium	Ba^{2+}	apple green	burns with a yellow-green flame to form oxide $2Ba + O_2 \rightarrow 2BaO$	reacts with cold water to form hydroxide and hydrogen $Ba + 2H_2O(l)$ $\rightarrow Ba(OH)_2 + H_2$	burns to form the white chloride salt $Ba + Cl_2 \rightarrow BaCl_2$

occurrence Calcium is the fifth most abundant element in the Earth's crust. *Vast quantities occur as the rocks* **chalk,** **limestone,** *and* **marble.** *These deposits are formed mainly from the shells of dead marine animals. Magnesium is the eighth most abundant element*

Mineral	Main chemical constituent
Limestone	$CaCO_3$ (calcium carbonate)
Chalk	$CaCO_3$ (calcium carbonate)
Marble	$CaCO_3$ (calcium carbonate)
Anhydrite	$CaSO_4$ (calcium sulphate)
Gypsum	$CaSO_4.2H_2O$ (calcium sulphate)
Magnesite	$MgCO_3$ (magnesium carbonate)
Dolomite	$CaCO_3.MgCO_3$ (mixed carbonate)

in the Earth's crust. It is found mainly as the carbonate (magnesite $MgCO_3$) often joined with calcium carbonate (dolomite). The other alkaline-earth metals are much rarer.

lime kiln A lime kiln is a heating tower used to make lime by the **thermal decomposition** of limestone. *The kiln is charged from the top with the limestone, and also with coke, which burns with the air to supply the heat to decompose the limestone. The carbon dioxide escapes from the top of the kiln and the lime falls, as a solid, to the bottom. Like many industrial processes, it is continuous.*

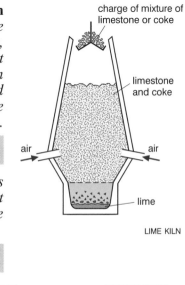

charge of mixture of limestone or coke

limestone and coke

air   air

lime

LIME KILN

> **limestone → lime + carbon dioxide**
> $CaCO_3 \rightarrow CaO + CO_2$

Lime is often called quicklime because it is 'quick' to react with water. When it does it has 'slaked' its thirst and forms slaked lime (calcium hydroxide).

> **lime + water → slaked lime**
> $CaO + H_2O \rightarrow Ca(OH)_2$

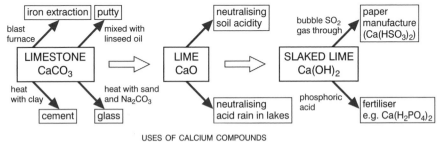

USES OF CALCIUM COMPOUNDS

uses of magnesium Magnesium is the only alkaline-earth metal used on a commercial scale: beryllium is too rare, and calcium, strontium, and barium are too reactive to have many uses. *Magnesium is combined with aluminium to make lightweight* **alloys.** *As magnesium burns with an intense white light, it is used as a light source in flares and in photography. Flash bulbs contain magnesium ribbon or powder in an atmosphere of oxygen.*

halogens (or **group VII elements**) are the elements in Group VII of the periodic table, which have seven valence electrons in their outermost shell. *The name is derived from Greek and means 'salt-makers'. Their ions and compounds are called **halides**.*

physical properties The halogens are all **diatomic molecules** (F_2, Cl_2, Br_2, and I_2) which have poisonous vapours or gases. *As we go down the group the molecules become larger and heavier, and the melting and boiling point increases. Fluorine and chlorine are gases, bromine is a liquid, and iodine is a solid (at room temperature and pressure).*

PHYSICAL PROPERTIES OF HALOGENS

Element	Diatomic molecule	Interatomic distance	Melting point (°C)	Boiling point (°C)	Solubility in water	Appearance
Fluorine	F_2	(F)——(F) 0.14 nm	−220	−188	very soluble	pale yellow-green gas
Chlorine	Cl_2	(Cl)——(Cl) 0.19 nm	−101	−35	quite soluble	pale green-yellow gas
Bromine	Br_2	(Br)——(Br) 0.23 nm	−7	59	slightly soluble brown vapour	dark red fuming liquid giving off
Iodine	I_2	(I)——(I) 0.27 nm	114	184	almost insoluble	silvery-black solid (gives off purple vapour on heating)

chemical properties The halogens are reactive **non-metals** whose reactivity decreases as we go down the group. *This is because as the atoms become larger it makes it more difficult for the nucleus to attract electrons. Because the halogens have seven valence electrons they react with metals to form the negatively charged **halide** ions F^-, Cl^-, Br^-, and I^-. As they are good at accepting electrons they act as **oxidising agents**.*

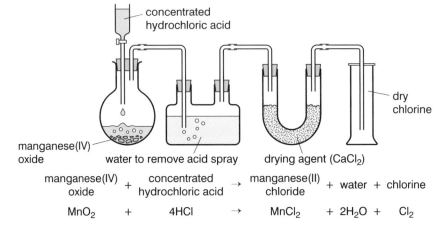

concentrated hydrochloric acid

dry chlorine

manganese(IV) oxide

water to remove acid spray

drying agent ($CaCl_2$)

manganese(IV) oxide	+	concentrated hydrochloric acid	→	manganese(II) chloride	+ water + chlorine
MnO_2	+	$4HCl$	→	$MnCl_2$	+ $2H_2O$ + Cl_2

MAKING CHLORINE GAS

- **Bleaching action** Halogens in aqueous solutions are bleaching agents. Chlorine dissolves in water to form chlorine water. This is a mixture of two acids, hydrochloric acid and chloric (I) acid.

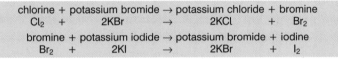

chlorine + water → hydrochloric acid + chloric(I) acid
Cl_2 + H_2O → HCl + HClO

Chloric (I) acid is a **bleach** as it is an **oxidising agent**. When the dye is oxidised it becomes colourless.

- **Displacement reactions** Any halogen above another in the group is more reactive and a more powerful oxidising agent. It will therefore displace any halogen below it from a solution of its salt. For example, chlorine displaces both bromide and iodide ions, as it oxidises these ions to the element.

chlorine + potassium bromide → potassium chloride + bromine
Cl_2 + 2KBr → 2KCl + Br_2

bromine + potassium iodide → potassium bromide + iodine
Br_2 + 2KI → 2KBr + I_2

- **Affinity for hydrogen** Chlorine has a strong attraction for hydrogen and it will remove hydrogen from many compounds, especially hydrocarbons, to form hydrogen chloride gas. Hydrogen will even burn in chlorine gas. This is used in the industrial preparation of **hydrochloric acid**, as hydrogen chloride gas readily dissolves in water to produce this acid.

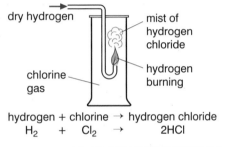

hydrogen + chlorine → hydrogen chloride
H_2 + Cl_2 → 2HCl

BURNING HYDROGEN IN CHLORINE GAS

halides are the salts formed when the halogens react with metals. *If chlorine gas is passed over heated iron wool, the iron wool glows brightly and forms a brown smoke of iron (III) chloride.*

testing for halides Metal halide salts are ionic substances and therefore usually dissolve in water. *However, the silver halides are insoluble and this can be used as a test for the halides.*

TESTING FOR HALIDES

Halide	Addition of silver nitrate solution
chloride	white precipitate of silver chloride AgCl
bromide	cream precipitate of silver bromide AgBr
iodide	yellow precipitate of silver iodide AgI

hydrogen halides are colourless gases which are very soluble in water and dissolve to form the corresponding acid.

Hydrogen halide	Acid formed
hydrogen chloride HCl(g)	hydrochloric acid HCl(aq)
hydrogen bromide HBr(g)	hydrobromic acid HBr(aq)
hydrogen iodide HI(g)	hydroiodic acid HI(aq)

uses of halogens

- **Fluorine** is added as fluoride salts to drinking water to reduce dental decay.
- **Chlorine** in solution is used as a bleach and for sterilising water.
- **Bromine** is used as bromide salts on black and white photography film.
- **Iodine** in solution in alcohol is used as a antiseptic (tincture of iodine). When dissolved in KI solution it is used as a test for **starch** (turns from brown to black).

planet Earth

the **shape of the Earth** is an oblate spheroid, which means that it is shaped like a ball but is slightly flattened at the poles. *The gravitational force is slightly greater at the poles than at the equator, because of the shorter distance to the centre of the Earth.*

latitude The latitude of a point on the Earth's surface is its distance from the equator, measured in degrees. *Points on the equator have a latitude of 0°. The North and South Poles have latitudes of 90°N and 90°S.*

parallel The parallels of **latitude** are imaginary circles drawn around the Earth, parallel to the equator.

longitude The longitude of a point on the Earth's surface is its distance around the Earth's circumference, measured in degrees east or west, and starting at the **meridian** which runs through Greenwich, near London. *Greenwich has a longitude of 0°.*

meridian The meridians of **longitude** are imaginary circles drawn around the Earth passing through both poles.

Earth's rotation The Earth rotates in an anticlockwise direction on its axis and one complete revolution takes 24 hours.

Earth's axis The Earth's axis is an imaginary line through the centre of the Earth. *The tilt of the Earth's axis explains why places at different latitudes have different lengths of day and night.*

day and **night** are caused by the rotation of the Earth on its axis: when a place on Earth faces the Sun it is day; when it faces away from the Sun it is night. *The time taken for the Earth to make one complete revolution on its axis is 24 hours and is called a solar day.*

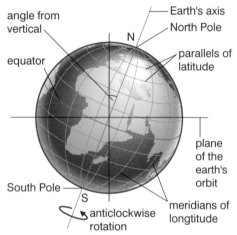

equinox An equinox is the time of year when both day and night are of equal lengths (12 hours each). *Equinoxes occur when the Sun appears to cross over the equator. This occurs twice a year. In the northern hemisphere the spring or vernal equinox is about March 21st and the autumn equinox about September 22nd. The dates are reversed for the southern hemisphere.*

solstice A solstice is the time of year when either the day or the night are at their longest. *In the northern hemisphere the summer solstice occurs on June 21st and the winter solstice on December 21st. The dates are reversed for the southern hemisphere.*

seasons are caused by the tilt of the Earth's axis as it orbits the Sun in an anticlockwise direction. *The half of the Earth which is tilted towards the*

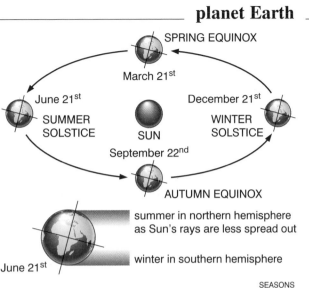

Sun will receive a greater concentration of Sun's energy per square metre. For example, on June 21st the northern hemisphere is tilted towards the Sun, so it is summer. In the southern hemisphere on June 21st, the Sun's energy is spread over a larger area, so it is winter. The situation is reversed on December 21st.

summer in northern hemisphere as Sun's rays are less spread out

winter in southern hemisphere

SEASONS

Moon The Moon is the Earth's only natural satellite, orbiting at a mean distance of 384 400 km from Earth. *It has a diameter of 3474 km and has no atmosphere or surface water. Its surface temperature varies between 80 K (night minimum) and 400 K (noon on the equator). The Moon takes 28 days to orbit the Earth. It is held in orbit by the gravitational attraction of the Earth.*

phases of the Moon are the eight different appearances of the Moon as it orbits the Earth in an anticlockwise direction. *As the Earth is rotating on its axis, the Moon appears to move quite quickly across the night sky.*

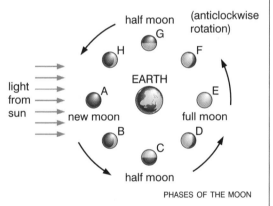

PHASES OF THE MOON

tides are the regular rise and fall of the water level in the Earth's oceans as a result of the gravitational forces between the Earth, Moon, and Sun. *The size of the gravitational force depends on both the distance and mass of the celestial body. Although the Moon is very much smaller than the Sun it is comparatively close. As a result, the Moon is approximately twice as effective as the Sun in causing tides.*

spring tides are extra high tides caused when both the Moon and the Sun are aligned with one another, increasing the gravitational force on the Earth's oceans.

neap tides are the relatively small tides caused when the Sun and Moon's gravitational forces are acting at right angles to one another, decreasing the overall gravitational force.

plants (roots)

roots are the lowest part of a plant, usually underground, whose functions are anchorage (secure fixing) and uptake of water and mineral salts. *Water is taken into the root by* **osmosis**.

structure of root The growing point (**meristem**) of the root is just behind the root tip. *This is protected by a root cap as it grows between the soil particles. In the piliferous (hairy) layer, which is the youngest growth area, there are long outgrowths from the outermost cells, called root hairs.*

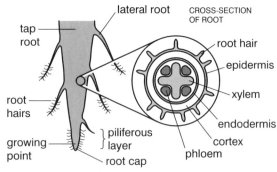

ROOT STRUCTURE

mineral salts are inorganic substances containing chemical elements essential to the health of the plant.

vascular tissue is the fluid-conducting tissue in the plant and consists

Element	Function in plant
Calcium	helps to hold cell walls together
Iron	part of enzyme which makes chlorophyll
Magnesium	part of chlorophyll molecule
Nitrogen	needed to make proteins
Phosphorus	essential for strong root system
Potassium	necessary for enzymes in photosynthesis

of two types: **xylem** and **phloem**. *In the roots of most flowering plants, the xylem tissue is arranged in an X-shaped mass and the phloem is found between the arms of the X.*

xylem The xylem is the **vascular tissue** of plants which allows water and dissolved **mineral salts** to move from the roots through the stem to the leaves. *The cells of the xylem are strengthened by a rigid substance called lignin in their cell walls. Xylem vessels form the woody part of a plant and do not contain living cytoplasm.*

phloem The phloem is the **vascular tissue** of plants which moves the products of **photosynthesis** from the leaves to the storage areas and growing points of a plant (**translocation**). *Phloem tissue consists of long columns of living cells (sieve tubes) which have lost their nuclei and cytoplasm and are connected by porous plates called 'sieve plates'.*

cambium The cambium is a thin layer of living **tissue** between the xylem on the inside and the phloem on the outside. *These cells are able to divide and make more xylem and phloem, and so form a* **meristem**.

meristem A meristem is a region of plant **tissue** which is actively dividing and producing living cells. *The most important meristems occur at the growing points at the top of the stem or shoot and at the tip of the root.*

vascular bundles (in seed plants) are regions of **vascular tissue** (called conducting tissue) which are responsible for the movement of water, mineral salts, and food from the leaves to the storage and growth organs. *Vascular bundles normally consist of* **xylem** *and* **phloem** *tissue separated by a living* **cambium** *layer. Roots often have a central vascular bundle, whereas stems have vascular bundles arranged in a ring near the outside edge.*

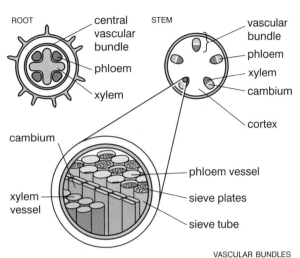

VASCULAR BUNDLES

root hair A root hair is an outgrowth from a single cell in the outer layer of the root, in the growth area behind the root tip. *Root hairs have a large surface area to volume ratio. They absorb water by* **osmosis** *and mineral salts by* **active transport**.

osmosis is the movement of a solvent (usually water) from a dilute to a more concentrated solution by **diffusion** across a **semipermeable membrane**. *The cytoplasm and cell sap inside the root are quite concentrated solutions and the water in the soil is a dilute solution. Water therefore diffuses into the root hair through the semipermeable cell membrane. A 'concentration gradient' allows the water to travel from cell to cell (dilute to more concentrated) until it reaches the* **xylem** *vessels in the centre of the root.*

semipermeable membrane (or **selectively permeable membrane**) This is a membrane which allows the molecules of water (solvent) to pass through, but not the molecules of most dissolved substances (solutes).

osmotic pressure is the pressure equal to that required to stop osmosis between a particular solution and pure water.

isotonic describes two solutions that have the same **osmotic pressure**. *Isotonic drinks have the same osmotic pressure as blood, so are quickly absorbed.*

hypertonic describes a solution with greater **osmotic pressure** than another.

hypotonic describes a solution with lower **osmotic pressure** than another.

active transport (or **active uptake**) is the movement of substances such as **mineral salts** through a membrane in living cells against a concentration gradient: i.e. from low to high concentration. *It is an energy-requiring process which occurs in the root hairs. The concentration of minerals in the soil is quite low in comparison to the concentration inside the root hair. Energy (supplied from the* **mitochondria** *of root hair cells) is used to move the mineral salts into the root hair. This is opposite to the direction they would normally take by diffusion.*

leaf A leaf is a structure of a plant, usually flat and green, which grows from the **stem**. *Green leaves produce food for the plant and are the sites for the processes of **transpiration** and **photosynthesis**. Simple leaves consist of a single leaf blade (lamina). Compound leaves have a number of small leaf blades (leaflets) growing from the same leaf stalk (petiole).*

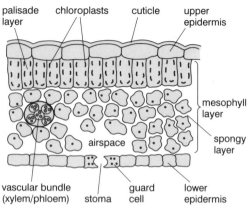

palisade layer   chloroplasts   cuticle   upper epidermis

mesophyll layer

spongy layer

airspace

vascular bundle (xylem/phloem)   stoma   guard cell   lower epidermis

TRANSVERSE SECTION OF A LEAF BLADE

leaf vein A leaf vein is a long strip formed by **vascular bundles** in the leaf. *They supply water and mineral salts and remove the food made in the leaf. Some plants like grasses have parallel veins, but most have a main vein (midrib) with branching veins coming off it.*

epidermis The epidermis is the layer of cells covering the top and bottom of a leaf.

cuticle The cuticle is a waterproof waxy coating secreted by the **epidermis** and protecting a leaf.

mesophyll layer The mesophyll layer is the middle layer of a leaf, between the upper and lower epidermis.

palisade mesophyll The palisade mesophyll is made up of regularly shaped cells which contain many **chloroplasts**.

spongy mesophyll The spongy mesophyll is made up of irregularly shaped cells with air spaces in between them, in which gases for photosynthesis can circulate by **diffusion**.

stoma (plural: **stomata**) A stoma is a pore found on the lower epidermis of a leaf, surrounded by a pair of **guard cells**. *These cells are responsible for opening and closing the stoma. The stoma opens when the guard cells become **turgid** and closes when they become **flaccid**. Stomata are responsible for letting gases in and out of the leaf during **photosynthesis**, and water vapour out by evaporation during **transpiration**.*

photosynthesis is the chemical process of separating hydrogen from water (light stage or photolysis) which then combines with carbon dioxide

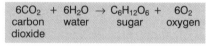

$$6CO_2 + 6H_2O \rightarrow C_6H_{12}O_6 + 6O_2$$

carbon dioxide   water   sugar   oxygen

(dark stage) to synthesise simple foodstuffs such as glucose. *It occurs in the **chloroplasts** of cells. Oxygen gas is released through the **stomata**.*

chlorophyll is a green pigment in the **chloroplasts** of plant cells and is the light-absorbing molecule needed for photosynthesis. *It absorbs the red and blue ends of the visible light spectrum and reflects green light.*

limiting factors are the main factors which can affect the rate of photosynthesis: light intensity (and wavelength), carbon dioxide concentration, and temperature. *Increasing each of these will increase photosynthesis, up to a certain maximum value. For example, low light intensity limits photosynthesis even if CO_2 concentration is high.*

transpiration is the process in which water is lost by evaporation from the leaves through the **stomata**. *The flow of water and mineral salts through the plant from the roots to the leaves is maintained by 'root pressure' (action of **osmosis** and **active transport**) and by the pull through the xylem vessels caused by water being evaporated from the leaves.*

turgor is inflation of a plant cell to a rigid state brought about by **osmosis**. *Healthy plant cells will take in no more water because the outward pressure (turgor pressure) equals the inward pressure of the cell wall.*

turgid describes the state of a cell which has full **turgor**.

flaccid describes the state of cells which are limp through the lack of **turgor**.

wilting is a state in which the plant is losing more water by **transpiration** than it can replace by **osmosis**, so that its cells become **flaccid**.

plasmolysis is an extreme state of **wilting** where a plant loses water not only by transpiration but also by 'reverse osmosis' caused by extremely dry conditions in the soil. *It may cause the death of the plant.*

potometer A potometer is an apparatus which measures **transpiration** rates under natural or artificial conditions. *High temperature, high light intensity (bright sunshine), low humidity, and windy conditions will all increase the rate of transpiration.*

translocation is the transport of minerals and products of photosynthesis within a plant. *Minerals travel in solution through the **xylem** vessels, and sugars travel in solution through the **phloem** vessels.*

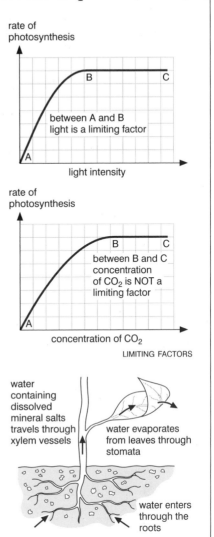

rate of photosynthesis

B   C

between A and B light is a limiting factor

A

light intensity

rate of photosynthesis

B   C

between B and C concentration of CO_2 is NOT a limiting factor

A

concentration of CO_2

LIMITING FACTORS

water containing dissolved mineral salts travels through xylem vessels

water evaporates from leaves through stomata

water enters through the roots

TRANSPIRATION STREAM

plants (stem and flowers)

stem A stem is the part of a plant which usually grows vertically upward towards the light. *Water and food travel up and down the stem through* ***vascular bundles*** *and it supports the leaves, buds, and flowers. Some plant species have organs formed from modified stems which are important in* ***vegetative reproduction:***

- **bulb** A bulb is a short, thick underground **stem** with scaly leaves containing stored food (e.g. daffodil, onion).
- **corm** A corm is a short, thick **stem** like a bulb but food is stored in the stem itself (e.g. crocus).
- **rhizome** A rhizome is a thick **stem** which grows horizontally underground and is used for food storage (e.g. lily, fern, iris).
- **stolon** (or **runner**) A stolon is a **stem** which grows horizontally from the bulb of a plant and forms roots for new plants where it touches the ground (e.g. strawberry).
- **tuber** A tuber is a swollen underground **stem** or **root** which acts as a food storage organ and produces buds for new plants (e.g. potato, dahlia).

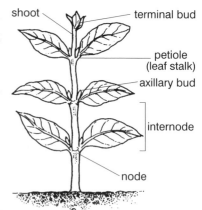

STEM ATTACHMENTS

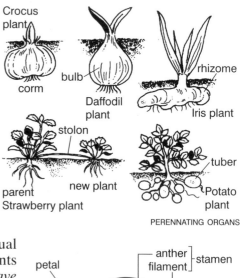

PERENNATING ORGANS

flower A flower is the organ of sexual reproduction in flowering plants (**angiosperms**). *Most flowers have both the male* ***stamens*** *and female* ***carpels*** *inside the same flower. Many flowers attract insects which help in* ***pollination***.

receptacle The receptacle is the expanded tip of the flower stalk which bears the **sepals**, **petals**, **stamens**, and **carpels**. *After fertilisation the receptacle may swell up to form a fleshy false fruit.*

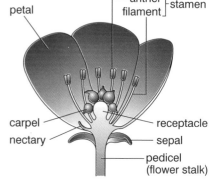

FLOWER STRUCTURE

sepal A sepal is a small protective leaf-like structure found around a flower bud. *When the flower opens, the sepals may fall off or remain as a ring underneath the petals.*

calyx is the collective name for all the **sepals**.

petal A petal is a flower structure which surrounds the reproductive organs (stamens and carpels) of the flower. *Petals are often brightly coloured and scented to attract insects.*

corolla is the collective name for all the **petals**.

nectary A nectary is an area of cells at the base of the petals which can secrete a sweet sugary liquid called 'nectar', which attracts insects for **pollination.**

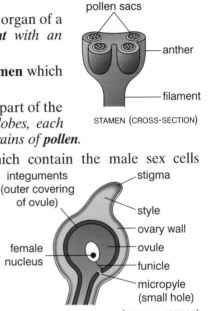

pollen sacs

anther

filament

STAMEN (CROSS-SECTION)

stamen A stamen is the male reproductive organ of a **flower**. *Typically each has a **filament** with an **anther** at its tip.*

filament A filament is the stalk of the **stamen** which supports the **anther**.

anther The anther is the pollen-producing part of the **stamen**. *It normally consists of two lobes, each containing two 'pollen sacs' with the grains of **pollen**.*

pollen is the yellow dust-like grains which contain the male sex cells (**gametes**) of the plant.

androecium is the collective term for all the male parts of a **flower**.

carpel (or **pistil**) A carpel is the female reproductive organ of a flower. *Typically each carpel has a **stigma**, **style**, and **ovary**. Some flowers have a single carpel, others have several clustered together.*

integuments (outer covering of ovule)

stigma

style

ovary wall

female nucleus

ovule

funicle

micropyle (small hole)

CARPEL (VERTICAL SECTION)

stigma The stigma is the uppermost part of the **carpel**. *During **pollination** it secretes a sticky substance so that pollen attaches to it.*

style The style is the part of the **carpel** between the stigma and the ovary. *The length of the style varies with the species of flower. The pollen tube grows through the style after **pollination**.*

ovary The ovary is the main reproductive structure of the **carpel** which contains one or more **ovules**.

ovule An ovule is the structure in female seed plants which contains the female sex cell (**gamete**). *It is attached to part of the ovary wall, and after fertilisation it develops into the seed.*

gynaecium is the collective term for all the female parts of a flower.

floral formula A floral formula shows the various numbers of the main parts of the flower for a particular species. *The main parts are the **calyx**, **corolla**, **androecium** and **gynaecium**. For example, $K_5C_5A_{10}G_{10}$ means 5 sepals, 5 petals, 10 stamens, and 10 carpels.*

inflorescence An inflorescence is a group of flowers on the same stalk.

pollution is an undesirable change in the environment as a direct result of human activities, both industrial and social. *Pollution may affect the atmosphere, the water (rivers, seas, oceans), or the land.*

pollutant A pollutant is any substance released into the environment as a result of human activities which has a harmful effect on living organisms. *Pollutants may be either **biodegradable** or **non-biodegradable**.*

biodegradable describes any substance which can be broken down by natural processes of decay. *Plant and animal waste (including **sewage**) are biodegradable and are broken down by the action of **bacteria** and **fungi**.*

non-biodegradable describes any substance which cannot be broken down by natural processes of decay. *Many plastics are non-biodegradable.*

air pollution (or **atmospheric pollution**) is the release into the atmosphere of toxic substances which have a harmful effect on the natural environment. *Most air pollutants are gases (or tiny smoke and lead particles) which are released into the **troposphere**.*

Air pollutant	Chemical formula	Source
carbon monoxide	CO	exhaust fumes
sulphur dioxide	SO_2	burning fossil fuels (coal, oil, natural gas)
CFCs	–	used in refrigerants, aerosols
oxides of nitrogen	$(NO)_x$	exhaust fumes
lead particles	Pb	exhaust fumes
smoke particles	C	coal and wood fires

smokeless zones are areas of towns and cities where only special smokeless fuels (not coal or wood) are allowed to be burnt, to reduce **air pollution**.

greenhouse effect The greenhouse effect is the trapping of heat energy in the atmosphere because of the effects of **greenhouse gases**. *The infrared radiation (heat energy) is given off from the Earth's surface as it is warmed up by the Sun.*

greenhouse gases are gases in the atmosphere which absorb infrared radiation, causing an increase in air temperature. *The most important is **carbon dioxide** which is increased by burning fossil fuels and by*

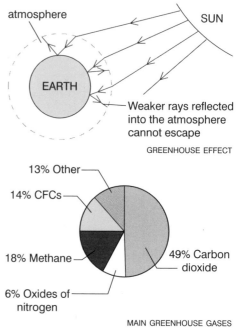

GREENHOUSE EFFECT

13% Other

14% CFCs

18% Methane

6% Oxides of nitrogen

49% Carbon dioxide

MAIN GREENHOUSE GASES

deforestation which reduces the amount of carbon dioxide removed by photosynthesis. Another is methane, a by-product of rice-farming and cattle-rearing.

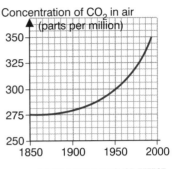

Concentration of CO$_2$ in air (parts per million)

GRAPH OF GREENHOUSE EFFECT

global warming is the gradual change in world climate caused by the **greenhouse effect**. *It is thought that it may cause changes in weather patterns, causing droughts and storms, and even melting the polar ice caps to bring about severe flooding.*

acid rain is rainwater which has a pH less than 5 due to dissolved gases such as sulphur dioxide and oxides of nitrogen. *These gases are produced mainly by burning fossil fuels. Normal rainwater has a pH of 5.6 due to dissolved carbon dioxide. With dissolved sulphur dioxide and oxides of nitrogen the pH can fall by up to 2 points, which is a 100-fold increase in acidity. Acid rain damages leaves, washes essential nutrients out of the soil, and kills fish and other aquatic life in lakes and rivers.*

lead pollution is caused by motor vehicles which emit tiny lead particles in their exhaust fumes. *Lead compounds are added to petrol so that it does not ignite too soon. Lead when inhaled can build up in the body, causing hyperactivity (especially in young children), liver and kidney damage, and a lowering of I.Q. Lead pollution is reduced by the use of unleaded petrol.*

ozone layer (or **ozonosphere**) The ozone layer is a region of the upper atmosphere at a height of between 20 and 40 km, containing a high concentration of **ozone** O$_3$. *The layer is formed from oxygen by ultraviolet radiation from the Sun, and absorbs ultraviolet radiation*

oxygen molecule O$_2$	$\xrightarrow[\text{light}]{\text{ultraviolet}}$	oxygen atoms 2O·
oxygen molecule O$_2$ + oxygen atom O·	$\longrightarrow$	ozone molecule O$_3$
ozone molecule O$_3$	$\xrightarrow[\text{light}]{\text{ultraviolet}}$	oxygen molecule O$_2$ + oxygen atom O·

which would otherwise reach the Earth's surface and cause harm to living organisms. The ozone layer is at risk from pollution by CFCs.

CFCs (abbreviation for **chlorofluorocarbons**) are inert chemicals used as refrigerants or as solvents in aerosols. *These CFCs float upwards into the upper atmosphere and react with solar radiation to release chlorine atoms which break down the protective ozone layer. 'Ozone holes' were first noticed over the South Pole in 1987. Increased amounts of ultraviolet radiation can kill tiny plants (phytoplankton) in the sea which are the main producers in many food chains. It can also cause skin cancer and eye cataracts in humans. Alternatives to CFCs are used in 'ozone-friendly' products.*

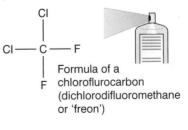

Formula of a chlorofluorocarbon (dichlorodifluoromethane or 'freon')

pollution II

water pollution results from human activities such as farming, industry, etc., causing various pollutants to dissolve in streams, rivers, and the sea. *Fertilisers used by farmers are washed from the soil into streams (**leaching**) and cause **eutrophication**. Other pollutants include agricultural waste (**slurry**), **effluent** and **sewage**, and **oil** from refineries and tankers.*

Water pollutant	Source
fertilisers	added to soil by farmers and leached from the soil by rainwater
effluent	industrial waste material including chemicals, solvents, detergents discharged from factories
sewage	this is human waste from washing and using the toilet. All sewage must be treated before being discharged into rivers or the sea
oil	from oil refineries or accidents at sea with oil tankers. Also many oil tankers wash out their holds after delivery

leaching is the washing out of soluble materials by a liquid passing through a solid. *It occurs when rainwater flows through the soil. Acid rain will leach away important minerals from the soil, and **fertilisers** are leached from the soil to cause **eutrophication** and **nitrate pollution**.*

eutrophication (or **unintentional enrichment**) is an overgrowth of aquatic plants like algae caused by the presence of dissolved fertilisers or untreated sewage which are rich in nitrates and phosphates. *These plants use up the oxygen in water, causing the death of fish and other aquatic life.*

effluent is waste material discharged from factories and other industrial sites, which may contain heavy metals such as mercury, lead, and cadmium, and solvents or **detergents**. *Strict laws ensure that before such waste is discharged into natural water it is treated to neutralise any harmful chemicals.*

oil pollution is caused by oil spillage into the sea, resulting in oil slicks which can devastate marine life and kill seabirds. *Originally oil spillages were treated with **detergents**, but this just spread out (dispersed) the oil even more. Nowadays it is thought best to contain the slick and allow natural bacteria in the water to feed on the oil **hydrocarbons** and break them down. To encourage the oil-eating bacteria to multiply, special fertilisers are spread over the oil slick. For small spillages, chemicals are sprayed on to the oil which solidify it to a rubbery material which can be 'rolled up'.*

thermal pollution is caused by warm water released from power stations and factories into rivers. *Warm water contains less dissolved oxygen and so is harmful to aquatic life.*

biological oxygen demand (abbreviation: **B.O.D.**) is a measure of the extent of pollution in water. *Polluting organic matter (agricultural waste, sewage, etc.) encourages the growth of microorganisms which use up the oxygen in water. High B.O.D. values indicate severe pollution.*

B.O.D. (mg dm^{-3})	Significance
below 30	no pollution
30–80	mild pollution
above 80	severe pollution

nitrate pollution is the pollution of water by nitrates, caused mainly by the **leaching** of fertilisers from the soil. *These dissolved nitrates can cause **eutrophication** of streams and rivers. They also cause health problems if they get into drinking water, as the body can convert them into nitrites, which can prevent the haemoglobin in the blood from absorbing oxygen. Babies are particularly vulnerable to this. The European Commission has set a maximum level of nitrates in drinking water of 50 mg/l.*

catalytic converter A catalytic converter is a device fitted to a car exhaust to convert harmful gases (carbon monoxide and oxides of nitrogen) into harmless gases (carbon dioxide and nitrogen). *The catalysts are usually platinum and rhodium. Other catalysts such as palladium will also oxidise unburnt hydrocarbon fuel to water and carbon dioxide.*

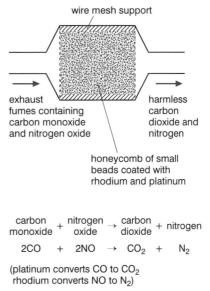

land pollution is a general term for pollution caused by the burying of household and commercial waste. *Waste material such as plastic is often **non-biodegradable** and remains in the soil for a long time. Radioactive waste (**nuclear waste**) from nuclear reactors and laboratories takes thousands of years to become safe.*

$$\underset{\text{monoxide}}{\text{carbon}} + \underset{\text{oxide}}{\text{nitrogen}} \rightarrow \underset{\text{dioxide}}{\text{carbon}} + \text{nitrogen}$$

$$2CO + 2NO \rightarrow CO_2 + N_2$$

(platinum converts CO to CO_2
rhodium converts NO to N_2)

CATALYTIC CONVERTOR

noise pollution is caused by excessive unpleasant sounds, typically loud or high-pitched. *Heavy urban traffic can produce noise levels of around 90 decibels. Large aircraft taking off produce around 120 decibels. Prolonged loud noise, such as over-amplified music or power tools, causes permanent hearing damage. There are many by-laws to minimise noise pollution. For example, car horns must not be used after dark, except in an emergency. Noise pollution can be reduced by fitting silencers to engines, mufflers to machinery, and sound insulation in buildings (curtains, carpets, acoustic tiles, double glazing, etc).*

electrostatic smoke precipitator An electrostatic smoke or dust precipitator is used in chimneys or extractor ducts to remove fine particles of soot or dust. *It does this by passing the smoke between charged plates, which induces a charge on the tiny particles and attracts them onto the plates. Periodically the electricity is switched off, and the dust falls off the plates and is collected.*

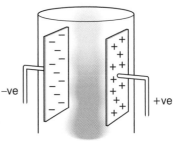

ELECTROSTATIC SMOKE PRECIPITATOR

polymer A polymer is a giant long-chained molecule made up of a large number of smaller molecules joined together.

synthetic polymers are substances such as **plastics**, and man-made fibres such as **nylon** and **Terylene**.

monomer A monomer is a small molecule that joins with others to form a polymer.

polymerisation is the **chemical reaction** by which monomers are joined together to form a polymer. *The number of monomers joining together may vary between 50 to 50 000 or more. Normally the reaction requires a particular range of temperature and pressure and the presence of a catalyst.*

addition polymerisation is an **addition reaction** between **monomers** which are **unsaturated molecules**.

polythene (or **polyethene**) is an **addition polymer** formed by up to 50 000 ethene molecules.

Part of a polythene molecule

high-density polyethene (abbreviation: **H.D.P.E.**) is formed when ethene gas is bubbled into a hydrocarbon solvent with special catalysts at $100°C$ and atmospheric pressure. *H.D.P.E. has few branching chains, so the polymer chains can pack close together. It is a hard plastic with a density around $0.96\ g\,cm^{-3}$. Its uses include making trays, bottle crates, etc.*

low-density polyethene (abbreviation: **L.D.P.E.**) is formed when ethene gas with a trace of oxygen gas (as an initiator) is heated to $200°C$ and subjected to extremely high pressures of 1500 atmospheres. *L.D.P.E.*

ADDITION POLYMERS

Monomer name	Monomer formula	Polymer name	Polymer formula	Uses
ethene	H₂C=CH₂	polyethene (polythene)	+(CH₂—CH₂)ₙ	1 plastic bags 2 plastic film 3 sheets
propene	CH₃CH=CH₂	polypropene (propathene)	+(CH(CH₃)—CH₂)ₙ	1 crates 2 rope 3 carpet
chloroethene (vinyl chloride)	CHCl=CH₂	polychloroethene (polyvinylchloride or P. V. C.)	+(CHCl—CH₂)ₙ	1 waterproof material 2 insulating wire 3 plastic records 4 guttering
phenylethene (styrene)	CH(C₆H₅)=CH₂	polyphenylethene (polystyrene)	+(CH(C₆H₅)—CH₂)ₙ	1 packaging material 2 ceiling tiles 3 insulating material
tetrafluro-ethene	F₂C=CF₂	polytetrafluroethene (PTFE or Teflon)	+(CF₂—CF₂)ₙ	1 coating for bridge bearings 2 non-stick saucepans 3 coating on skis
methyl methacrylate	C(COOCH₃)(F)=C(F)(F)	polymethyl-methacrylate (Perspex)	+(C(COOCH₃)—CH₂)ₙ	1 substitute for glass

has many branching chains, which cannot be packed together tightly. It is a soft plastic with a density around 0.92 g cm^{-3}. Its uses include making plastic bags, bin liners, cling film, and packaging.

condensation polymerisation is the successive linking together of **monomers** to form a polymer with the elimination of a simple molecule such as water or hydrogen chloride. *A polymer may contain two kinds of monomer.*

nylon (or **polyamide**) is a **synthetic polymer** made by **condensation polymerisation** of a diamine monomer and a dicarboxylic acid monomer.

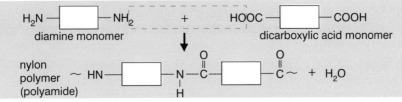

Terylene (or **polyester**) is a **synthetic polymer** made by **condensation polymerisation** of a diol monomer with a dicarboxylic acid chloride monomer.

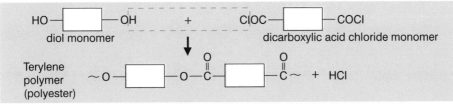

plastics are a class of materials which when subjected to heat and pressure become soft so they can be easily shaped. *Most plastics are synthetic polymers although a few like cellulose are natural polymers. Plastics are relatively cheap, lightweight, good insulators, easy to clean, translucent (see-through) or easily coloured, non-corrosive, long lasting, and can be very strong. However, they are often difficult to dispose of because they are non-biodegradable, and when they burn they often produce toxic fumes.*

thermoplastic describes a type of **plastic** which can be resoftened on heating. *Thermoplastics include addition polymers like polythene and polystyrene which can be easily moulded and shaped. This is because there are only weak forces between the polymer chains.*

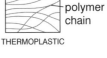

thermosetting describes a type of **plastic** which on heating becomes permanently hard as strong cross-linking develops between the polymer chains. *Examples include formica, bakelite, and melamine.*

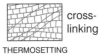

moulding plastics Soft or molten plastic can be moulded by compression, injection into a mould, or extrusion. Hollow objects like plastic bottles and dolls can be made by 'blow moulding'.

polymers (natural)

natural polymers are **polymers** produced by living organisms. *They include natural rubber and natural foodstuffs. All natural polymers are biodegradable, unlike most synthetic polymers.*

macromolecules are extremely large molecules. *Natural polymers are also known as macromolecules.*

polysaccharides are **natural polymers** formed by **condensation polymerisation** of simple sugars, often **glucose**.

glucose ($C_6H_{12}O_6$) is a simple **sugar** which is the main **monomer** unit in most **polysaccharides**, and provides the fuel for **respiration** in cells.

glycogen (or **animal starch**) is a **polysaccharide** found in vertebrate animals and is the main energy store in the liver and muscles. *It is converted by the enzyme **amylase** into glucose which is used in **respiration** to release energy.*

cellulose is a **polysaccharide** which acts as the main structural material of plants, found in their **cell walls**. *It consists of very long unbranched chains of glucose monomers.*

starch is a **polysaccharide** found in plants, especially in the roots, tubers, seeds, and fruit. *It is an important **carbohydrate** energy source. It is formed by the **condensation polymerisation** of around 3000 **glucose** units.*

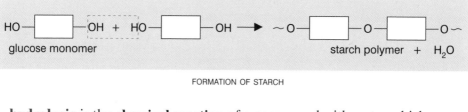

glucose monomer   starch polymer + H_2O

FORMATION OF STARCH

hydrolysis is the **chemical reaction** of a compound with water which causes it to break down. *A polymer which undergoes hydrolysis breaks into smaller pieces, or right down into **monomers**.*

starch hydrolysis is the **hydrolysis** of starch into smaller molecules, and eventually into **glucose**. *It is important in the digestion of starchy foods.*

- **enzyme hydrolysis** is **starch hydrolysis** by the enzyme **amylase**, found in saliva in the mouth. It breaks the starch down into the disaccharide maltose $C_{12}H_{22}O_{11}$ which contains two glucose units minus a water molecule.

$$\text{starch} + \text{water} \xrightarrow{\text{amylase}} \text{maltose}$$
$$(C_6H_{10}O_5)_x + \tfrac{x}{2}H_2O \longrightarrow \tfrac{x}{2}C_{12}H_{22}O_{11}$$

- **acid hydrolysis** is **starch hydrolysis** by an acid and takes place in the stomach of mammals. Acid hydrolysis is slow but eventually the starch is broken down into glucose, which is the **monomer** and will not undergo further hydrolysis.

$$\text{starch} + \text{water} \xrightarrow{\text{acid}} \text{glucose}$$
$$(C_6H_{10}O_5)_x + xH_2O \longrightarrow xC_6H_{12}O_6$$

fats are naturally occurring **esters** of a fatty acid and glycerol. *They are solids at body temperature and are widely used by plants and animals as a means of storing food which can be used as a fuel.*

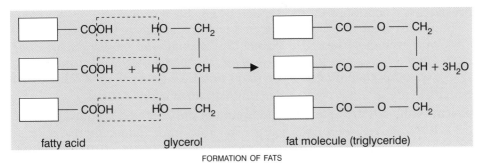

fatty acid   glycerol   fat molecule (triglyceride)

FORMATION OF FATS

hydrolysis of fats (or **saponification**) is the **chemical reaction** of a fat with water to break it down into a **fatty acid** and glycerol. *Hydrolysis of fats is important in digestion, and in the manufacture of **detergents**.*

proteins are **natural polymers** of **amino acids** which form all enzymes, and also the main structural materials of animals. *Proteins contain the elements carbon, hydrogen, oxygen, nitrogen, and sulphur. (Fats and carbohydrates only contain carbon, hydrogen, and oxygen.) Their molecular mass may vary from about 6000 to several million.*

globular proteins are proteins which have compact rounded molecules and are usually water-soluble. *Examples of globular proteins are **enzymes**, **antibodies**, **haemoglobin**, some **hormones**, and storage proteins such as casein in milk and albumin in egg-white.*

fibrous proteins are proteins which are generally insoluble in water and consist of long coiled strands or flat sheets which give strength and elasticity for structure. *Examples of fibrous proteins are actin and myosin in **muscle**, keratin in **skin** and hair, and collagen in **bone**.*

amino acids are the monomer units of all **proteins** and contain the –COOH and –NH$_2$ groups at either end of the molecule. *Amino acids undergo **condensation polymerisation** by eliminating a water molecule which forms the **peptide link** between the monomers.*

peptide links are the –CO–NH– linkage formed between amino acids during protein synthesis. *Many of these links form a peptide chain. Peptide chains join with one another to form polypeptides, which again join to form **proteins**.*

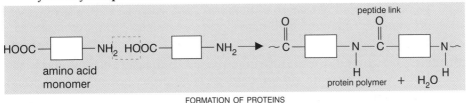

amino acid
monomer   protein polymer + H$_2$O

FORMATION OF PROTEINS

hydrolysis of proteins is the **chemical reaction** of protein molecules with water to produce **amino acids**. *It is important in protein digestion and takes place in the stomach and small intestine. Hydrolysis of proteins should not be confused with **denaturing** (partial breakdown by heat).*

pressure is a continuous force applied by an object or fluid against a surface, measured as the force acting per $$pressure = \frac{force}{area}$$ unit area of surface. *The greater the force and the smaller the area, the larger the pressure. It is because of pressure that a sharp knife (small area) cuts better than a blunt knife (large area), and a snow shoe (large area) does not sink into snow like an ordinary shoe (small area).*

pascal (symbol: **Pa**) A pascal is the SI unit of pressure and is equivalent to a force of 1 newton acting over an area of 1 square metre: $1\,Pa = 1\,N\,m^{-2}$. *It was named after the French physicist Blaise Pascal (1623–62).*

atmospheric pressure is the pressure exerted by the air and is caused by the gravitational attraction of the air to the Earth. *It acts equally in all directions and decreases with altitude. Atmospheric pressure is measured using bars ($1\,bar = 10^{5}\,Pa$) or millibars ($1\,mb = 100\,Pa$).*

barometer A barometer is an instrument which measures atmospheric pressure.

mercury barometer A mercury barometer is a simple barometer which consists of a glass tube about 80 cm long which is completely filled with mercury. *It is inverted and the open end is submerged in a dish of mercury. The mercury in the tube drops until the pressure of the mercury at the base of the tube is equal to the air pressure acting on the mercury in the dish. At sea level, atmospheric pressure is about 760 mm of mercury (equivalent to 101 325 Pa).*

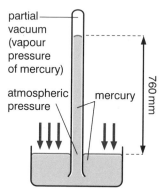

MERCURY BAROMETER

aneroid barometer An aneroid barometer is a type of barometer formed from a metal box with a thin corrugated lid. *The air inside this box is removed and the lid is supported by a spring. Changes in pressure cause the lid to move against the spring. This movement is magnified with levers to control a pointer. Aneroid barometers are not as accurate as mercury barometers, but are more robust.*

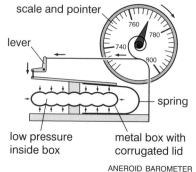

ANEROID BAROMETER

altimeter An altimeter is an instrument used to measure height above sea level, e.g. in aircraft. *Atmospheric pressure decreases with height, so an **aneroid barometer** can be used as an altimeter.*

isobar An isobar is a line joining places with the same atmospheric pressure. *Isobars are used on weather maps, which use the millibar (mb) as a unit of pressure. In the UK, atmospheric pressure varies from 975 mb (low pressure) to 1030 mb (high pressure).*

depression (or **cyclone** or **low**)
A depression is a region of low atmospheric pressure caused when air flows in from below and rises to flow out at a higher level. *As the air is rising upward it cools. It therefore holds less water vapour, so depressions are associated with wet weather and rain.*

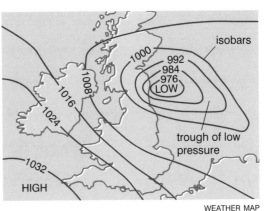

WEATHER MAP

anticyclone An anticyclone (or a **high**) is a region of high atmospheric pressure caused when air flows in from above. *As it descends it becomes warmer and therefore can hold more water vapour. Clouds therefore disperse, and so anticyclones are associated with dry weather.*

manometer A manometer is a U-shaped tube with a liquid inside which is used for measuring fluid (gas or liquid) pressure. *The pressure to be measured is fed in to one side of the tube and the other is left open to the atmosphere. The difference in level of the liquid in the two limbs gives a measure of the unknown pressure.*

gas pressure = atmospheric pressure + pressure of liquid (shown by $h\rho g$)

MANOMETER

siphon A siphon is an inverted U-tube with one end longer than the other, which moves a liquid from one place to another place at a lower level. *For the siphon to work, the tube must be full of liquid before it is placed in position.*

pressure in liquids Gravity acting on a liquid causes pressure to be exerted on the walls of its container. *This, like air pressure, acts equally in all directions but does not depend on the shape of the container. It does depend on the density of the liquid and it increases with depth. The pressure at a point in a liquid is given by:*

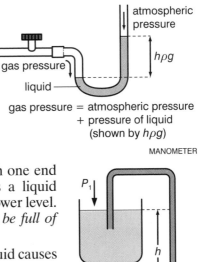

$$P_1 = P_2 + h\rho g$$

SIPHON

pressure in liquid at a point	=	height(h) of liquid at point	×	density (ρ) of liquid	×	gravitational constant (g)

This is in addition to the atmospheric pressure above the liquid. In water, the pressure increases by approximately one atmosphere for every 10 metres of depth. This is why the walls of a dam are much thicker at the bottom.

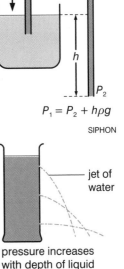

pressure increases with depth of liquid

radioactivity is the spontaneous disintegration of unstable atomic nuclei and is usually accompanied by the emission of **radiation**. *This may be in the form of a stream of **alpha particles, beta particles,** or **gamma rays**.* *Radioactivity is an entirely random process and you* Symbol for *cannot predict which nuclei will decay next. It is completely unaffected by physical conditions (such as temperature) or chemical conditions (such as bonding).*

Symbol for radioactive hazard

radiation is a general term applied to anything that travels outward from its source but which cannot be identified as a type of matter like a solid, liquid, or gas. *Radiation applies to forms of **energy, electromagnetic waves,** and **radioactivity.***

radioisotope (or **radioactive isotope**) A radioisotope is an isotope of an element that is radioactive. *Some radioisotopes are produced by **cosmic rays**, others by **nuclear fusion**. Carbon-14 is a natural radioisotope formed from nitrogen by bombardment with cosmic rays, and present in small proportions among ordinary carbon-12 atoms.*

natural radioactivity (or **background radiation**) is the result of spontaneous disintegration of naturally occurring **radioisotopes** found in rocks and living material. *It increases underground because of the surrounding rocks, and at high altitudes because of the effect of cosmic rays. Local areas may have high natural radioactivity because the rocks below (e.g. igneous rocks such as granite) give off radon gas.*

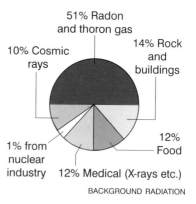

51% Radon and thoron gas

10% Cosmic rays

14% Rock and buildings

12% Food

1% from nuclear industry

12% Medical (X-rays etc.)

BACKGROUND RADIATION

background count The background count is a measure of the **natural radioactivity** in a particular place.

alpha particle (symbol: α) An alpha particle is a positively-charged helium nucleus which is ejected from certain radioactive nuclei. *It is a relatively heavy particle (with 4 times the mass of a proton) and so alpha radiation is the least penetrating form. However, because it is positively charged it attracts electrons from nearby atoms. It is therefore strongly ionising.*

alpha radiation is formed by a stream of **alpha particles**.

beta particle (symbol: β) A beta particle is a high-energy electron emitted from certain radioactive nuclei. *These particles can travel almost at the speed of light. They are much lighter than alpha particles and are more penetrating but have less ionising effect.*

beta radiation is formed by a stream of **beta particles** (electrons) emitted by certain radioactive nuclei.

CHARACTERISTICS OF RADIATION

Type	Nature	Charge	Radioactive source (e.g.)	Absorbed by	Ionising power	Deflection in electric and magnetic field	HAZARD
Alpha ^4_2He	helium nucleus (2 protons and 2 neutrons)	+2	americium-241	sheet of paper	strong	very small	low unless within body
Beta $^0_{-1}e$	high energy electron	−1	strontium-90	5 mm sheet of aluminium	weak	large	can damage cells and DNA
Gamma γ	high energy electro-magnetic wave	none	cobalt-60	25 mm sheet of lead (reduces intensity by half)	very weak	zero	dangerous in high intensity

gamma ray (symbol: γ) A gamma ray is an **electromagnetic wave** of very short wavelength and high frequency. *Such rays are often emitted at the same time as an alpha or beta particle. They have no charge or mass and have the least ionising effect of nuclear radiations but they are the most penetrating, and potentially the most hazardous.*

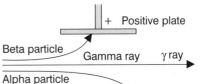

DEFLECTION IN AN ELECTRIC FIELD

radioactive decay is the spontaneous disintegration of a radioactive nucleus, giving off alpha or beta particles, often together with gamma rays. *If the new nucleus formed (daughter nucleus) is also radioactive, then this also undergoes disintegration and a radioactive decay series results.*

$$^{235}_{92}U \xrightarrow{\alpha} {}^{231}_{90}Th \xrightarrow{\beta} {}^{231}_{91}Pa \xrightarrow{\alpha} {}^{227}_{89}Ac$$
$$\downarrow \beta$$
$$^{215}_{84}Po \xleftarrow{\alpha} {}^{219}_{86}Rn \xleftarrow{\alpha} {}^{223}_{88}Ra \xleftarrow{\alpha} {}^{227}_{90}Th$$
$$\downarrow \alpha$$
$$^{211}_{82}Pb \xrightarrow{\beta} {}^{211}_{83}Bi \xrightarrow{\alpha} {}^{207}_{81}Tl \xrightarrow{\beta} {}^{207}_{82}Pb$$

RADIOACTIVE DECAY SERIES

half-life (symbol: $T_{\frac{1}{2}}$) The half-life is the time taken for half the atoms in a radioactive sample to undergo **radioactive decay**. *Half-life is therefore the time for the radiation emitted to be halved in intensity. Half-life values vary enormously. For example, radon gas has a half-life of 51.5 seconds, whereas carbon-14 has a half-life of 5600 years.*

Disintegrations per second

Background radiation is included in this curve. It must be subtracted to work out the true half-life.

Activity

40
20
10
5
0

51.5   103   154.5   206

$T_{\frac{1}{2}}$ $T_{\frac{1}{2}}$ $T_{\frac{1}{2}}$

Time (s)

DECAY OF RADON GAS

alpha decay is the **radioactive decay** of a nucleus giving off an **alpha particle**. *The remaining nucleus (new element) has its **atomic number** decreased by two and its **mass number** decreased by four.*

beta decay is the **radioactive decay** of a nucleus by conversion of a neutron into a proton, giving off a **beta particle** (high-energy electron). *The remaining nucleus (new element) has its **atomic number** increased by one but its **mass number** remains the same.*

gamma emission is caused by a rearrangement of particles inside the nucleus of the atom, resulting in a loss of energy as gamma rays. *By itself, it causes no change in **atomic number** or **mass number**.*

nuclear equation A nuclear equation shows the **nuclides** and **radiation** involved in a nuclear reaction.

nuclide A nuclide is an atomic nucleus as defined by its **atomic number** and **mass number**.

alpha decay	$^{235}_{92}U \rightarrow {}^{231}_{90}Th + {}^{4}_{2}He$	(alpha particle)
beta decay	$^{14}_{6}C \rightarrow {}^{14}_{7}N + {}^{0}_{-1}e$	(beta particle)
gamma decay	$^{1}_{0}n + {}^{238}_{92}U \rightarrow {}^{239}_{92}U + \gamma$	(gamma ray)

radionuclide A radionuclide is a nuclide which is radioactive and decays to give off alpha particles, beta particles, gamma rays, or a combination of these.

detecting radiation Most devices detect and measure radioactivity by detecting the amount of ionisation it causes.

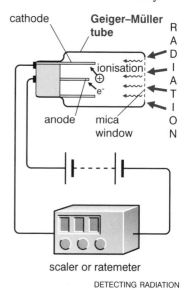

DETECTING RADIATION

- **Geiger counter** A Geiger counter is an electronic device which detects the charge produced when a particle causes ionisation, and indicates the number of particles detected in the tube per second.
- **cloud chamber** A small enclosed chamber is saturated with alcohol vapour and the radioactive source placed inside. Movement of alpha and beta particles causes ionisation which results in condensation tracts.
- **bubble chamber** Like a cloud chamber, but shows particle tracks as tiny bubbles caused by ionisation.
- **scintillation counter** Detects gamma rays which, when they hit a special crystal, produce flashes called 'scintilla' of light.
- **dosimeter (film badge)** Simple device containing photographic film which darkens on exposure to radiation. The larger the dose, the darker the film.

becquerel (symbol: **Bq**) A becquerel is the SI unit for measuring radioactivity, equal to the activity in a material in which one nucleus decays on average per second.

gray (symbol: **Gy**) A gray is the SI unit for measuring absorbed dose of radiation, equivalent to the absorption of $1\,J\,kg^{-1}$ of ionising radiation.

sievert (symbol: Sv) A sievert is the SI unit for measuring dose equivalent, equal to grays multiplied by a factor depending on the type of radiation and the kind of tissue being irradiated.

handling radioactive sources In school laboratories, radioactive sources should have activities of around 3.7 kBq or less. *All sources must be handled at arm's length with tweezers and work should not be undertaken by anyone having a cut, abrasion, or open wound. Outside the body, both beta and gamma sources are dangerous. Inside the body, alpha particles (e.g. from inhaled radioactive gases) also cause tissue damage.*

radiation sickness Radiation sickness is illness caused by exposure to large doses of radiation. *Such exposure can result in various forms of cancer, especially leukaemia (blood cancer). Gamma rays especially can damage cells and cause genetic **mutation**. Workers using radioisotopes wear protective clothing such as lead-lined aprons, and use tools such as long tweezers. Hazardous radioisotopes should be handled in a **glove box**.*

glove box A glove box is a closed cupboard that has gloves fixed into holes in the wall of the cupboard. *Glove boxes for highly radioactive material may be shielded by concrete and special glass so that an operator can safely handle highly radioactive material and toxic chemicals.*

radiotherapy is the use of radiation from **radioisotopes** to treat cancer by killing cancer cells.

radiography is the process of producing images of opaque objects on photographic film or fluorescent screen using radiation such as gamma rays or X-rays.

radiology is the study of radioactivity, and especially the use of gamma rays and X-rays in medical diagnosis and treatment.

radioactive tracing (or **radioactive labelling**) is a method of following substances by introducing a radioactive isotope, called the tracer. *Tracers in underground gas or water pipes can be used to detect leaks. They are used in medicine to follow the movement of body fluids like blood, urine, etc. Such tracers are gamma sources with short half-lives.*

carbon dating By comparing the amounts of carbon-14 in dead material (like wooden artefacts, leather sandals, etc.) with the levels of carbon-14 in living material, we can measure the age of the dead material. *It takes 5600 years (the half-life of carbon-14) for the activity to halve. The proportion of carbon-14 in living material is kept constant, as food eaten contains carbon-14. On death this ceases, so the level of carbon-14 starts to fall. This method of carbon dating assumes that the levels of carbon-14 in living material has not changed over thousands of years.*

rock dating By measuring the amount of radioactive isotope left in a rock sample, and knowing the half-life of the isotope, you can calculate how old the rock is.

irradiation (or **sterilisation**) is the process of exposing something to radiation. *It is used to kill bacteria in food, sterilise hospital equipment, etc.*

rates of reaction

rate of reaction is the speed of a **chemical reaction**, calculated by measuring how quickly **reactants** (chemicals you start with) change into **products** (chemicals you produce). *The rate of a particular chemical reaction is affected by various factors like the temperature, concentration, and surface area of the reactants, or the presence of a catalyst.*

measuring rates To follow a particular **chemical reaction** we need to be able to see and record some sort of change. *Not all chemical reactions are easy to follow. Some are too quick, others have no visible change. Changes can most easily be observed:*
- when there is a change in mass as a gas is given off.
- when a gas is formed which can be collected.
- when there is a temperature change.
- when there is a colour change.
- when a precipitate is formed.
- when there is a change in pH.

rate curve A rate curve is typically a plot of change of mass or volume during the reaction against time. *To start with the gradient of the graph is steep as the rate of reaction is always fastest at the beginning. As the reactants are used up the gradient decreases until the reaction has finished and the curve becomes flat.*

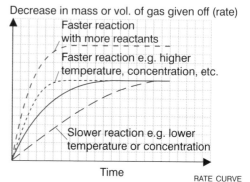

Decrease in mass or vol. of gas given off (rate)

Faster reaction with more reactants

Faster reaction e.g. higher temperature, concentration, etc.

Slower reaction e.g. lower temperature or concentration

Time

RATE CURVE

collision theory explains rates of reaction in terms of the motion of particles in the reactants.
- **temperature** At a higher temperature, reactant particles are moving faster with greater average kinetic energy. A greater propotion of them therefore collide with enough energy to be converted from reactants to products. Rates of reaction which are slow at room temperature often double with a 10 degree rise in temperature.
- **concentration** At a higher concentration, there is a greater chance of reactant particles colliding with each other with enough energy to be converted into products. Rate of reaction therefore doubles if concentration is doubled.
- **pressure** (in gases) Increasing pressure decreases the volume of a certain mass of gas. This means that there are more particles in a certain volume (increase in concentration). Therefore there is a greater likelihood of collision and a faster rate of reaction at higher pressures.
- **surface area** Smaller particles, e.g. in powders, have a much greater surface area than lumps or crystals. With a greater surface area, more collisions can take place. Rate of reaction therefore doubles if the surface area of the reactant particles doubles.

activation energy is the amount of energy which colliding particles must have in order to start a chemical reaction and change reactant particles into products. *This energy is used in the breaking of chemical bonds.*

Catalyst	Process	Chemical reaction
Iron (Fe)	Haber process	nitrogen + hydrogen $\xrightarrow{\text{Fe}}$ ammonia
Vanadium (V) oxide	Contact process	sulphur dioxide $\xrightarrow{\text{V}_2\text{O}_5}$ sulphur trioxide
Nickel (Ni)	Hydrogenation of margarine	unsaturated hydrocarbon $\xrightarrow{\text{Ni}}$ saturated hydrocarbon
Platinum (Pt)	} Catalytic converters	carbon monoxide $\xrightarrow{\text{Pt}}$ carbon dioxide
Palladium (Pd)		hydrocarbons $\xrightarrow{\text{Pd}}$ water + carbon dioxide
Rhodium (Rh)		nitrogen oxides $\xrightarrow{\text{Rh}}$ nitrogen

catalyst A catalyst is a substance that increases the rate of a **chemical reaction** without itself undergoing any permanent chemical change. *A catalyst is often very specific and may increase the rate of one particular reaction but have no effect on another reaction (see* **enzymes***). Most industrial catalysts are* **transition metals** *or their compounds. A catalyst works by lowering the* **activation energy**.

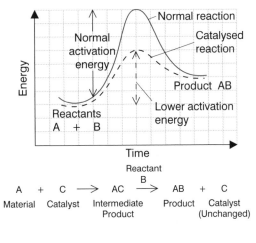

$$A + C \longrightarrow AC \xrightarrow{\text{Reactant B}} AB + C$$

Material   Catalyst   Intermediate   Product   Catalyst
Product   (Unchanged)

ACTION OF CATALYSTS

catalysis is the process of changing the rate of reaction by using a **catalyst**.

intermediate An intermediate is a short-lived chemical species sometimes formed in a chemical reaction and stabilised by the presence of a catalyst. *This intermediate is still very unstable, and reacts to form products and the original catalyst. The physical appearance of the catalyst may change but it does not change chemically.*

surface catalyst A surface catalyst holds reactant particles on its surface for long enough so that they can react with each other to form product(s). *Many surface catalysts are transition metals which speed up the reaction between gases.*

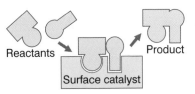

Reactants   Product

Surface catalyst

promoter (or **activator**) A promoter is a substance which increases the power of a catalyst and thereby speeds up the chemical reaction even more.

inhibitor (or **negative catalyst**) An inhibitor is a substance which slows down a chemical reaction, often by reducing the power of the catalyst. *Inhibitors sometimes 'poison' the catalysts by destroying intermediate chemical species.*

reactivity series of metals

reactivity series (or **activity series**) The reactivity series is a list of metals placed in order of their reactivity, as determined by their reaction with air (oxygen), water and dilute acid. *The most reactive metals are at the top of the series and the least reactive at the bottom. The element **hydrogen** is sometimes placed in the series for comparison. Any element below hydrogen can never displace hydrogen from water or an acid.*

reaction of metals with oxygen Reactivity of metals with oxygen (air) decreases as we go down the **reactivity series**. *Metals at the top of the series react violently and catch fire. The further down the series the metal is, the slower its reaction. If there is a reaction, the product is always an **oxide**.*

REACTION OF METALS WITH OXYGEN

Metal	Reaction with oxygen (air)	Chemical equation
Potassium	Violent reaction catching fire and burning with a lilac flame	$4K + O_2 \rightarrow 2K_2O$
Sodium	Easily catches fire and burns with a bright yellow flame	$4Na + O_2 \rightarrow 2Na_2O$
Magnesium	Can be ignited to burn with a brilliant white light	$2Mg + O_2 \rightarrow 2MgO$
Iron	Does not burn but glows brightly and gives off yellow sparks	$3Fe + 2O_2 \rightarrow Fe_3O_4$
Copper	Does not burn but the hot metal becomes coated with a black oxide	$2Cu + O_2 \rightarrow 2CuO$

reaction of metals with water Reactivity of metals with water decreases as we go down the **reactivity series**. *Metals at the top of the series react violently with cold water to produce the **alkali** and **hydrogen**. Metals near the middle react with steam to produce the **oxide** and **hydrogen**. Metals like copper have no reaction and never displace hydrogen.*

REACTION OF METALS WITH WATER

Metal	Reaction with water	Chemical equation
Potassium	Very violent reaction and metal catches fire	$2K + 2H_2O \rightarrow 2KOH + H_2$
Sodium	Violent reaction in cold water	$2Na + 2H_2O \rightarrow 2NaOH + H_2$
Calcium	Steady reaction in cold water	$Ca + 2H_2O \rightarrow Ca(OH)_2 + H_2$
Magnesium	Reacts very slowly with cold water but vigorously with steam	$Mg + H_2O \rightarrow MgO + H_2$
Zinc	Quite slow reaction with steam	$Zn + H_2O \rightarrow ZnO + H_2$
Iron	Very slow reaction with steam	$3Fe + 4H_2O \rightarrow Fe_3O_4 + 4H_2$
Copper	No reaction at all	

reaction of metals with hydrochloric acid Reactivity of metals with an acid decreases as we go down the **reactivity series**. *Metals at the top of the series (sodium and potassium) should never be reacted with acids as their reaction is too violent. As we go down the reactivity series, the reaction becomes slower, until with lead we need concentrated hydrochloric acid for any reaction. The reaction produces a **salt** called a chloride and **hydrogen gas**. For metals like copper and those below, there* is no reaction with hydrochloric acid and no hydrogen gas evolves. They will react with concentrated sulphuric or nitric acids, but will not produce hydrogen gas.

REACTION OF METALS WITH HYDROCHLORIC ACID

Metal	Reaction with hydrochloric acid	Chemical equation
Magnesium	Violent reaction	$Mg + 2HCl \rightarrow MgCl_2 + H_2$
Zinc	Moderate reaction	$Zn + 2HCl \rightarrow ZnCl_2 + H_2$
Iron	Slow reaction	$Fe + 2HCl \rightarrow FeCl_2 + H_2$
Lead	Reaction only when concentrated acid is used	$Pb + 2HCl \rightarrow PbCl_2 + H_2$
Copper	No reaction at all	
Gold	No reaction at all	

displacement reaction (or **substitution reaction**) A displacement reaction is a **chemical reaction** where a more reactive element (metal higher in the series) replaces a less reactive element (metal lower in the series) from its compound. *For example, if an iron nail is left in blue copper(II) sulphate solution, the iron displaces the copper. The blue colour is replaced by the pale-green colour of iron(II) sulphate solution, and pink copper metal is deposited on the iron nail.*

iron	+	copper(II) sulphate	$\rightarrow$	iron(II) sulphate	+	copper
Fe	+	$CuSO_4$	$\rightarrow$	$FeSO_4$	+	Cu

stability of metal compounds Metals high in the **reactivity series** form stable compounds whereas compounds of metals low down in the series are less stable and are easier to decompose.

STABILITY OF METAL COMPOUNDS

Metal	Carbonate	Nitrate	Hydroxide
Potassium Sodium	Do not decompose on heating	Decompose on heating to **nitrite** and **oxygen**	Do not decompose on heating
Calcium Magnesium Zinc Iron Copper	Decompose on heating to **oxide** and **carbon dioxide**	Decompose on heating to **oxide**, **nitrogen dioxide** and **oxygen**	Decompose on heating to **oxide** and **steam**
Silver	Decomposes above 100°C	Decomposes to **metal**, nitrogen dioxide, and oxygen	Decomposes to **metal** and **steam**

CARBONATE   copper(II) carbonate $\xrightarrow{heat}$ copper(II) oxide + carbon dioxide
$CuCO_3 \longrightarrow CuO + CO_2$

NITRATE   potassium nitrate $\xrightarrow{heat}$ potassium nitrite + oxygen
$2KNO_3 \longrightarrow 2KNO_2 + O_2$

NITRATE   zinc nitrate $\xrightarrow{heat}$ zinc oxide + nitrogen dioxide + oxygen
$2Zn(NO_3)_2 \longrightarrow 2ZnO + 4NO_2 + O_2$

HYDROXIDE   magnesium hydroxide $\xrightarrow{heat}$ magnesium oxide + steam
$Mg(OH)_2 \longrightarrow MgO + H_2O$

vegetative reproduction (or **vegetative propagation**) is **asexual reproduction** in which a part of the parent plant (called a perennating organ) is able to develop into a new plant. *Examples of perennating organs are **bulb, corm, rhizome, stolon** (runner), and **tuber**.*

clone A clone is a genetically identical descendent produced by vegetative reproduction from an original plant seedling. *The new bulbs (called bulbils) which grow off the side of an old daffodil bulb are clones. The layered stolon of a strawberry plant is a clone of its parent. As these clones are genetically identical to their parent plant, any disease which affects the parent will also affect the clone.*

micropropagation is the taking of cuttings of the stem of a parent plant, each with a new bud. *Each cutting is sterilised and placed in a special growing medium containing growth hormones. The stem grows roots, and the bud develops into a new plant which is a **clone** of the parent.*

tissue culture is the growth of groups of plant cells (normally **meristem** tissue) away from the parent plant in a suitable culture medium containing growth hormones. *Growth is very rapid and can produce disease-free plants all year round. However, the plants are **clones**, whose lack of genetic variation make them vulnerable to new diseases.*

sexual reproduction involves the joining of male and female **gametes** (sex cells). *In flowering plants the male gamete is found in the **pollen** and the female gamete in the **ovule**.*

pollination is the transference of **pollen** from the **anther** (male reproductive organ) to the **stigma** (sticky part of female reproductive organ).

self-pollination occurs when the pollen is transferred from the anther to the stigma of the same flower.

COMPARISON OF INSECT AND WIND POLLINATION

Insect pollination (e.g. wallflower)	Wind pollination (e.g. grass)
usually strongly scented	no scent
usually have nectaries at base of the flower	no nectaries
large petals, often with 'guide-lines'	small inconspicuous petals or none at all
anthers and stigma inside the flower so insect has to brush past them to get to the nectar	anthers and stigma dangle outside flower so pollen can catch wind and stick to large stigma
sticky or spiky pollen to stick to insects	smooth, light pollen to be blown in the wind (causes hayfever)
small quantities of pollen	large amounts of pollen as most is lost in the wind

cross-pollination is when the pollen is transferred from the anther to the stigma of a different flower of the same species. *The pollen has to be carried by a pollinating agent such as the wind, insects, birds, water, etc.*

fertilisation is the fusion of the male nucleus (from the **pollen**) with the female nucleus (in the **ovule**). *This happens after pollination. The pollen,* *if it lands on the right type of **stigma**, grows a pollen tube down through the **style** towards the **ovary**. It does this by secreting enzymes to digest a path through the style. The male nucleus travels down the pollen tube. In the ovary each ovule is protected by a double layer of cells (integuments). At one end there is a small hole (micropyle). The male nucleus passes through this hole and fuses with the female nucleus in the ovule. Fertilised ovules become **seeds**.*

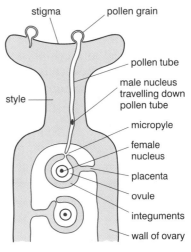

FERTILISATION IN A WALLFLOWER

seed A seed is a fertilised ovule together with its store of food. *When the two gametes join together during fertilisation, a **zygote** is formed. This divides by **mitosis** to form an **embryo** plant. This embryo and its store of food is the seed. Most seeds have the following parts:*

- **hilum** A scar on the seed showing where the **ovule** was attached to the **ovary**.
- **testa** Protective seed coat developed from the integuments.
- **cotyledon (seed leaf)**. A store of food (starch and protein) for the developing plant. Cotyledons also contain **enzymes**.
- **plumule** Primary bud inside the seed that will grow into the shoot of the new plant.
- **radicle** This will grow into the first root of the new plant.

fruit True fruit develops from the ovary wall after fertilisation. *The **ovary** wall forms a protective layer (pericarp) around the seed. Examples are the stone of a plum, the hard shell of a nut, and the pod of a bean. In a 'false fruit', the **receptacle** of the old flower swells up to form a fleshy outer core, as in apples, pears, or strawberries. These false fruits help in **seed dispersal** as they are attractive to eat.*

seed dispersal is the scattering of seeds from a plant over a wide area to avoid overcrowding and competition from other plants. *For wind dispersal, seeds need to be small and light, or have shaped fruits to carry them in the wind like the 'wings' of a sycamore or the 'parachute' of the dandelion. Seeds for animal dispersal may have hooked fruits to catch on animal fur, or succulent fruits and berries to be eaten by animals and birds. The seeds have a hard shell so they pass through an animal's digestive system unharmed. Some plants rely on self-dispersal in which the seeds are ejected fom the dried seed cases.*

germination is the initial stages of growth of a seed to form a seedling. *In order to germinate, seeds require warmth, moisture, and oxygen. During germination the **plumule** (young shoot) and **radicle** (young root) emerge through the seed coat. A seed will often remain 'dormant' until conditions are suitable for germination.*

reproduction (animal)

asexual reproduction is reproduction that involves only one parent and produces offspring which are genetically identical to their parent.

binary fission is a method of **asexual reproduction** in which the genetic material and cytoplasm of a single-celled organism divides equally to form two new cells. *These 'daughter cells' are genetically identical to the parent cell and are therefore **clones**. Bacteria under favourable conditions multiply in this way every 20 minutes. Within less than 10 hours, one bacterium could multiply to over a million.*

budding is a method of **asexual reproduction** used by simple animals like hydra, which grow a new group of cells out of the side of the parent body. *This 'bud' breaks off to form a new individual. In colonial animals such as coral, the new individual remains attached though it is self-contained. Budding also occurs in certain fungi such as **yeast**.*

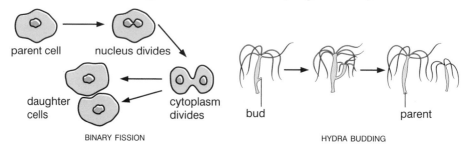

BINARY FISSION   HYDRA BUDDING

parthenogenesis is a method of **asexual reproduction** found in animals such as some **insects** (bees, aphids) where the development of a new organism is from an unfertilised egg. *The eggs formed by the female during parthenogenesis contain the full (diploid) number of chromosomes and are genetically identical.*

sexual reproduction is a type of reproduction which involves the fusion of specialised male and female sex cells called **gametes**. *This process is called fertilisation. Normally in animals it requires two parents,*

Asexual reprodution	Sexual reproduction
one parent	two parents
no gametes	gametes
identical genes	exchange of genes
cell division by mitosis	meiosis forms gametes
identical offspring	variation in offspring

but in plants both male and female reproductive organs are often on the same parent. Sexual reproduction involves the mixing of genetic material from both parents. This helps to give vigour and variation to a species.

sperm (or **spermatozoon**; plural: **spermatozoa**) is the male **gamete** of animals. *In male mammals it is produced in the **testis**. It consists of a head containing the genetic material and a tail (flagellum) for movement.*

ovum (plural: **ova**) or **egg** An ovum is the female **gamete** of animals. *It is produced in the **ovary**. It is normally larger than the sperm and is not mobile. It is spherical, with a nucleus and cytoplasm. The cytoplasm acts as a 'food store' after **fertilisation**.*

gamete (or **sex cell**) A gamete is a specialised sex cell formed by **meiosis** which contains only half the number of chromosomes (**haploid**). *In animals the gametes are the* **sperm** *(male) and* **ovum** *(female). Gametes combine during* **fertilisation** *to form the* **zygote** *which develops into the new offspring.*

gonads are reproductive organs of an animal. *These normally occur in pairs, two* **testes** *in the male and two* **ovaries** *in the female. Gonads also act as* **endocrine glands** *which secrete many important* **sex hormones**.

fertilisation is the fusion of the nuclei of the male and female **gametes** during sexual reproduction to form a single cell called the zygote. *Fertilisation normally takes place in the* **oviduct** *of the female. The* **sperm** *(male gamete) reaches the* **ovum** *(female gamete) by swimming (using its tail) from the vagina through the uterus and meets the egg travelling along the oviduct. Usually only one sperm enters the egg. Only the head containing the nucleus goes in, leaving the tail outside. When the successful sperm has entered the egg, the egg membrane becomes hard so no other sperm can enter. The unsuccessful sperm will all die.*

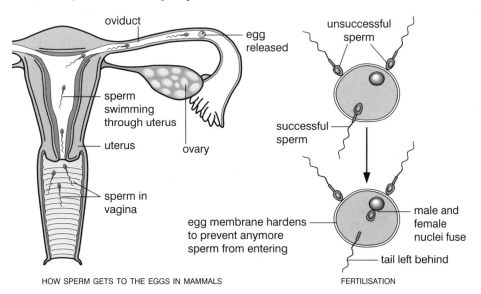

external fertilisation is fertilisation which occurs outside the body of the female. *It occurs in most aquatic animals.*

internal fertilisation is fertilisation which occurs inside the body of the female, into which sperm are introduced. *It occurs in most terrestrial animals, including humans.*

respiration is the release of energy in a living organism which occurs when simple products are made from the breaking down of food molecules. *Respiration occurs in all living cells all the time (cellular respiration).*

aerobic respiration is respiration which uses oxygen (air) and which releases energy and produces carbon dioxide and water. *These products are similar to combustion products, and this type of respiration is sometimes referred to as 'slow combustion'.*

glucose	+	oxygen	$\rightarrow$	carbon dioxide	+	water
$C_6H_{12}O_6$	+	$6O_2$	$\rightarrow$	$6CO_2$	+	$6H_2O$

anaerobic respiration is respiration which takes place in the absence of oxygen (air) and in which food substances are only partially broken down. *It produces **lactic acid** or **alcohol** and releases smaller amounts of energy than aerobic respiration. In humans, anaerobic respiration often occurs in the muscles during vigorous exercise when not enough oxygen is available.*

glucose	$\rightarrow$	lactic acid	$\rightarrow$	ethanol	+	carbon dioxide
$C_6H_{12}O_6$	$\rightarrow$	$2CH_3CHOHCOOH$	$\rightarrow$	$2C_2H_5OH$	+	$2CO_2$

oxygen debt occurs when the body requires more oxygen than is available (e.g. during vigorous exercise). *As a result, **anaerobic respiration** results in a build up of lactic acid, especially in the muscles. Deep breathing after the vigorous exercise repays this debt and oxidises the lactic acid eventually to carbon dioxide and water. A good measure of fitness is how quickly you can return to normal breathing (your 'recovery time').*

gaseous exchange (or **external respiration**) involves the passing of oxygen and carbon dioxide in and out of an organism. *This often takes place in a respiratory **organ** with a large surface area, e.g. **lungs** (adult amphibians, reptiles, birds, and mammals), gills (larval amphibians, fish and other aquatic animals), trachea and spiracles (insects) or **stomata** (plants).*

respiratory surface is the part of the respiratory organ through which **gaseous exchange** takes place. *This surface should be thin so that gases can diffuse through quickly and have a large surface area to speed up the gaseous exchange. It is also kept moist to improve diffusion, and has a rich supply of vessels.*

lung A lung is an organ for **gaseous exchange** in vertebrates.

epiglottis is a flap of cartilage which closes off the trachea and prevents food from going into the lungs when you swallow.

trachea (windpipe) is a tube through which air is drawn into the lungs.

larynx (voice box) is a region of the trachea containing the vocal cords, which vibrate to produce sound.

bronchus (plural: **bronchi**) is a branch of the trachea leading to a lung.

bronchiole is a terminal air tube in the lungs.

alveoli (singular: **alveolus**) are the tiny air sacs at the end of each bronchiole through which oxygen diffuses in and carbon dioxide diffuses out.

pleural membrane (or **pleura**) is a double membrane around the lungs filled with fluid which lubricates the lungs as they move against the rib cage.

mucous membrane has **goblet cells** which secrete **mucus** to lubricate the respiratory tract and trap germs. *Tiny hairs called **cilia** beat to move the mucus towards the mouth where it is swallowed.*

diaphragm (**midriff**) is a sheet of muscle situated below the lungs.

intercostal muscles are muscles between the ribs.

breathing (or **ventilation**) is the movement of air in and out of the lungs. *It involves the physical movement of the **diaphragm** and **intercostal muscles**. A relaxed adult has a breathing rate of about 16–18 times per minute.*

inspiration (or **inhalation**) is the movement of air into the lungs by the **diaphragm** being pulled down (flattened) and the rib cage being raised by contraction of the **intercostal muscles**.

expiration (or **exhalation**) is the movement of air out of the lungs by the **diaphragm** relaxing upwards and the **intercostal muscles** relaxing so that the rib cage moves inwards.

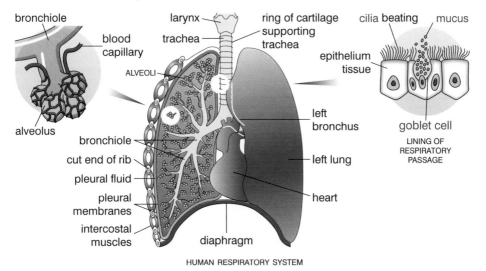

HUMAN RESPIRATORY SYSTEM

lung capacity The capacity of an adult human lung is about 5 litres. *In normal breathing only about 500 cm^3 of air is breathed in and out (tidal volume). However, there is no stagnant air, as inspired air mixes with*

Gas	Approximate volume	
	inhaled	exhaled
oxygen	21%	17%
nitrogen	78%	78%
carbon dioxide	0.03%	4%

all the air inside the lung. Both oxygen and carbon dioxide gases are soluble in the mucus of the lungs. Nitrogen gas is insoluble at normal atmospheric pressure, so the percentage of nitrogen in inhaled and exhaled air remains the same.

reversible reactions and equilibria

reversible reaction A reversible reaction is a **chemical reaction** in which the products react with one another to reform the original reactants, which then react to form products, and so on. *Eventually a* **chemical equilibrium** *is reached.*

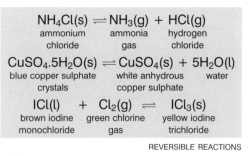

$$NH_4Cl(s) \rightleftharpoons NH_3(g) + HCl(g)$$
ammonium   ammonia   hydrogen
chloride   gas   chloride

$$CuSO_4.5H_2O(s) \rightleftharpoons CuSO_4(s) + 5H_2O(l)$$
blue copper sulphate   white anhydrous   water
crystals   copper sulphate

$$ICl(l) + Cl_2(g) \rightleftharpoons ICl_3(s)$$
brown iodine   green chlorine   yellow iodine
monochloride   gas   trichloride

REVERSIBLE REACTIONS

reversible sign The sign $\rightleftharpoons$ is used in the formulae of reversible reactions.

forward reaction The forward reaction is the direction from original reactants to products in a reversible chemical reaction. *It goes from left to right in the* **chemical equation***.*

backward reaction The backward reaction is the direction from products back to reactants in a reversible chemical reaction. *It goes from right to left in the* **chemical equation***.*

irreversible reaction An irreversible reaction is a **chemical reaction** which continues until one or all of the reactants is used up. *The products do not react with each other and the reaction goes to completion.*

reversible process is a term which commonly applies to physical changes which can be reversed by a change in conditions like temperature and pressure. *Reversible processes include melting, boiling, dissolving, etc. If the process is a chemical reaction then it is referred to as a* **reversible reaction***.*

open system An open system is one in which materials and energy can escape or enter. *If the products from a reversible reaction escape (e.g. as gases) then this destroys the reversibility. A chemical equilibrium cannot be established in an open system.*

closed system A closed system is one in which no material can escape, though energy can. *The temperature may not remain constant, so that no chemical equilibrium can be established.*

isolated system An isolated system is one in which no material or energy can escape or enter. *If the reaction is reversible this allows a* **chemical equilibrium** *to be established, at a particular temperature.*

static equilibrium is a type of **equilibrium** where a balance is reached so there is no movement of material. *Two people balanced on a see-saw provide an example of static equilibrium.*

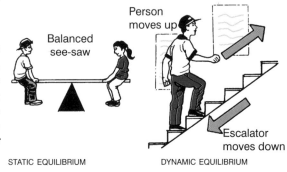

Balanced see-saw

STATIC EQUILIBRIUM

Person moves up

Escalator moves down

DYNAMIC EQUILIBRIUM

dynamic equilibrium is a type of **equilibrium** where a balance is reached because movement in one direction is cancelled out by the same rate of movement in the other direction. *A person walking up a moving escalator at the same rate as the escalator is going down is an example of dynamic equilibrium. Overall the person stays in the same position.*

phase equilibrium is a type of **dynamic equilibrium** established between phases or states of matter. *At a particular temperature a liquid will establish an equilibrium with its vapour. At this point the same number of particles leave from and return to the surface of the liquid (evaporate and condense).*

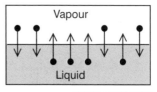

PHASE EQUILIBRIUM

chemical equilibrium is a stage reached in a reversible chemical reaction when the forward and backward reactions take place at the same rate. *This means that the overall concentration of reactants and products remain the same. A chemical equilibrium can only occur in an **isolated system** and is a **dynamic equilibrium**.*

Le Chatelier's principle states that if a change in temperature, pressure or concentration occurs to a system in **chemical equilibrium**, then the system will tend to adjust itself so as to counteract the effect of that change so that a new chemical equilibrium is established. *Consider the **Haber process** which makes ammonia from nitrogen and hydrogen gases.*

- **temperature** *The forward reaction gives out heat and is **exothermic**. Therefore, according to Le Chatelier's principle, if the temperature is lowered, more ammonia is produced. This counteracts the effect of a decrease in temperature. The backward reaction takes in heat and is **endothermic**. Therefore if the temperature increases, more nitrogen and hydrogen gases are produced. This counteracts the effect of the increase in temperature. Low temperature favours ammonia production (but not too low or the rate will be too slow: optimum is 450°C).*

- **concentration** *If we increase the concentration of the reactants, hydrogen and nitrogen, then according to Le Chatelier's principle, the equilibrium will move to the right-hand side. This will produce more ammonia, and counteract the effects of the increase in concentration of hydrogen and nitrogen. High concentration favours ammonia production.*

- **pressure** *High pressure would favour the side of the equilibrium which has the lowest volume. This would counteract the effect of an increase in pressure. This is the ammonia side, as 2 volumes of ammonia are produced from 4 volumes (1 volume of nitrogen and 3 volumes of hydrogen) of reactants. High pressure (normally around 250 atmospheres) favours ammonia production.*

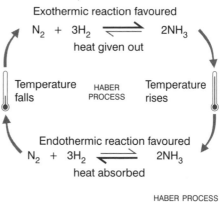

HABER PROCESS

salts are **chemical compounds** formed when the hydrogen of an acid is partially or wholly replaced by a metal or other positive ion. *Typically salts have **ionic bonds** and are normally formed by the reaction of an **acid** and **base**.*

ACID	+	BASE	→	SALT	+	WATER
sulphuric acid	+	copper oxide	→	copper sulphate	+	water
$H_2SO_4(aq)$	+	$CuO(s)$	→	$CuSO_4(aq)$	+	$H_2O(l)$

acid radical An acid radical is a group of atoms in an acid which becomes negatively charged when in solution. *Acid radicals cannot exist by themselves but can join with metal ions to form **salts**.*

nomenclature of salts If an acid ends in '–ic' then the salt has an ending '–ate' (except with hydrochloric acid) but if the acid has an ending '–ous' (in the old nomenclature) then the salt has an ending '–ite'.

Acid	Salt name (acid radical)
hydrochloric	chloride Cl^-
sulphuric	sulphate SO_4^{2-}
sulphurous	sulphite SO_3^{2-}
nitric	nitrate NO_3^-
nitrous	nitrite NO_2^-
phosphoric	phosphate PO_4^{3-}
phosphorous	phosphite PO_3^{3-}
carbonic	carbonate CO_3^{3-}
methanoic	methanoate $HCOO^-$
ethanoic	ethanoate CH_3COO^-

water of crystallisation is water that is present in definite proportions in a crystal, which is normally a **salt**. *Such salts are called hydrated salts, and the chemical formula shows the number of molecules of water of crystallisation associated with each molecule of the hydrate: e.g. hydrated copper(II) sulphate $CuSO_4.5H_2O$. This water of crystallisation can usually be removed by heating.*

anhydrous describes salts which have lost their **water of crystallisation**.

efflorescence is a process by which some hydrated salts lose their water of crystallisation. *Washing soda (hydrated sodium carbonate $Na_2CO_3.10H_2O$) if left exposed to the air is efflorescent and forms a white powdery deposit of **anhydrous** sodium carbonate Na_2CO_3.*

deliquescence is the spontaneous absorption by a substance of water from the atmosphere. *Anhydrous salts are often deliquescent, and some absorb so much water that a concentrated solution of the salt is formed.*

preparation of soluble salts There are four ways to prepare soluble salts.

metal	+	acid	→	salt	+	hydrogen		
metal oxide	+	acid	→	salt	+	water		
metal carbonate	+	acid	→	salt	+	water	+	carbon dioxide
metal hydroxide	+	acid	→	salt	+	water		

neutralisation is the chemical reaction of a **base** and an **acid** to produce a **salt** and water. *Everyday examples include the use of indigestion tablets containing a base such as MgO to neutralise excess hydrochloric acid in the stomach, and of lime (the base CaO) to neutralise acidic soil.*

titration is a technique which can be used for **neutralisation** of an acid and an alkali. *The alkali is accurately measured using a **pipette** and placed in a conical flask with an **indicator**. The acid is then added slowly from a **burette**. When the indicator changes colour, the **neutral** point has been reached.*

burette A burette is a long vertical graduated glass tube with a tap at one end which is used in **titration** to add controlled volumes of liquids.

pipette A pipette is a graduated glass tube which is filled by suction and used for transferring exact volumes ($10\,cm^3$, $25\,cm^3$) of liquids.

preparation of insoluble salts involves the reaction between two dissolved ionic substances (e.g. salts or acids) which 'change partners' to form a new insoluble salt (and a soluble salt or acid).

Soluble	Insoluble
all nitrates	(none)
most chlorides	Ag, Pb, Hg chlorides
most sulphates	Ca, Ba, Pb sulphates
Na, K, NH_4 carbonates	most carbonates

SOLUBILITY OF SALTS

double decomposition (or **ionic precipitation**) is a reaction producing **insoluble salts** by swopping the ionic partners of two soluble salts.

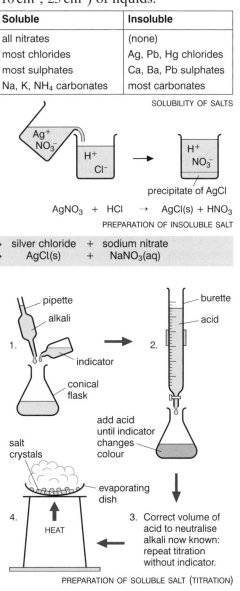

precipitate of AgCl

$$AgNO_3 \;+\; HCl \;\rightarrow\; AgCl(s) + HNO_3$$

PREPARATION OF INSOLUBLE SALT

silver nitrate	+	sodium chloride	→	silver chloride	+	sodium nitrate
$AgNO_3(aq)$	+	$NaCl(aq)$	→	$AgCl(s)$	+	$NaNO_3(aq)$

sodium chloride (or **common salt** or **table salt** or **rock salt** or **sea salt**) is the commonest and most important salt with a chemical formula of NaCl. *It is extracted from sea water, which contains about 2.7% by mass, by leaving shallow pools of sea water to evaporate in the sun. It can also be extracted from underground deposits of rock salt by pumping water underground to dissolve the salt. The concentrated salt solution (**brine**) is then pumped to the surface and left to evaporate. Sodium chloride is used for preserving and seasoning food and for de-icing roads.*

PREPARATION OF SOLUBLE SALT (TITRATION)

brine is a concentrated solution of **sodium chloride** (common salt).

scientists and a few of their achievements

c.3000 BCE	Invention of the **wheel** in Asia.
c.700 BCE	Invention of the **pulley**.
c.200 BCE	**Archimedes** (287–212) invented a **screw** for raising water.
c.1000	Chinese discovered how to make **gunpowder** and used it to propel missiles and rockets.
c.1100	Chinese first to use a **magnetic compass**.
c.1440	German Johannes **Gutenberg** (1400–68) invented first moveable type **printing press**.
c.1520	German Joseph **Kotter** invented the **rifle**.
c.1600	The Italian scientist **Galileo Galilei** invented the **pendulum**, **thermometer**, and **astronomical telescope**.
1615	English physician William **Harvey** (1578–1637) discovered the **circulation of the blood** in the human body.
1662	Irish scientist Robert **Boyle** (1627–1691) investigated gases and invented the **compressed air pump**.
1665	Experimental physicist Robert **Hooke** (1635–1703) first to invent the **compound lens** for a microscope.
1665–7	English scientist Isaac **Newton** (1642–1727) derived **laws of gravitation** and **motion** and investigated **colour** and **light**.
1750	Swedish biologist Carl **Linne** (**Linnaeus**) (1707–78) proposed a system of **biological classification** which is still used today.
1765	Scottish engineer James **Watt** (1736–1819) invented the high-pressure **steam engine**.
1796	English physician Edward **Jenner** (1749–1823) used **inoculation** to prevent people catching smallpox.
1800	Italian Count Alessandro **Volta** (1745–1827) devised the first **voltaic cell** by dipping different metals in brine.
1803	English chemist John **Dalton** (1766–1844) postulated that elements consisted of tiny indivisible particles (**atoms**).
1806–8	Englishman Sir Humphrey **Davy** (1778–1829) extracted various **metals** such as barium, calcium, magnesium, potassium, and sodium.
1811	Italian Count Amadeo **Avogadro** (1776–1856) investigated gases and concluded equal volumes have the same number of particles (**Avogadro constant**).
1821–31	English scientist Michael **Faraday** (1791–1867) discovered **electromagnetism** and used this to invent the electric **motor**, **transformer**, and **dynamo**.
1827	British botanist Robert **Brown** (1773–1850) first identified the random motion of small solid particles suspended in a liquid or gas (**Brownian motion**).

1859	British naturalist Charles **Darwin** (1809–82) put forward a theory of **evolution** by natural selection.
1866	Austrian monk Gregor **Mendel** (1822–84) formulated two important laws about **hereditary characteristics**.
1869	Russian chemist Dimitri **Mendeleev** (1834–1907) devised the modern **periodic table** of elements.
1875	French microbiologist Louis **Pasteur** (1822–95) showed that **microbes** were responsible for disease.
1876	Scottish-American inventor Alexander Graham **Bell** (1847–1922) invented the **telephone**.
1879	American Thomas Alva **Edison** (1847–1931) invented the **electric light bulb** and **phonograph** (gramophone).
1895	Italian inventor Guglielimo **Marconi** (1874–1937) invented wireless telegraphy (**radio**).
1897	English scientist Sir Joseph **Thomson** (1856–1940) first discovered negative particles, later called **electrons**.
1898	Polish scientist Marie **Curie** (1867–1934) investigated **radioactive decay** and isolated the elements polonium and radium.
1905	German born physicist Albert **Einstein** (1879–1955) published the first part of his theory of **relativity**.
1919	Ernest **Rutherford** (1871–1937) was the first scientist to 'split' the **atom** and identify the **proton**.
1926	Scotsman Logie **Baird** (1888–1946) demonstrated transmission of the first **television** pictures.
1928	Scottish bacteriologist Alexander **Fleming** (1881–1955) discovered the **penicillin** mould which killed bacteria.
1946	An early electronic machine (**computer**) called ENIAC (electronic numerical integrator and calculator) was built.
1961	English biochemist Francis **Crick** (1916–) and American biochemist James **Watson** (1928–) helped identify the structure of the **DNA** molecule.
1969	American Neil **Armstrong** (1930–) landed on the Moon.
1971	American Ted **Hoff** invented the **microprocessor**.
1978	The first **test-tube baby** was born.
1986	The Soviet space station **Mir** was put into orbit.
1998	Ian **Wilmut** and Keith **Campbell** made a **clone** of a mammal from an adult cell. The animal was a sheep called **Dolly**.
2000	Scientists produce a rough draft of the entire human **genetic code**.
2003	Discovery of a possible tenth planet in our **solar system**.

sensitivity (or **irritability**) is the characteristic property of all living organisms to be able to detect, interpret, and respond to changes in their environment. *Multicellular animals have specialised sense organs like the ears and eyes and effector organs like muscles and glands. Simple unicellular organisms like amoeba have no nervous system and the reception and response to a stimulus occurs in the same cell.*

photoperiodism is an organism's response to the length of day or night. *Often this initiates an important event in the life-cycle of the organism. This event is often linked with reproduction, ensuring that it occurs in the right season: e.g. plants flowering in spring, or birds migrating in winter.*

taxis (or **tactic movement**) is the movement of an organism with respect to a stimulus from a specific direction. *This stimulus may be chemical (chemotaxis), light (phototaxis), gravity (geotaxis), etc. For example, woodlice show negative phototaxis (move to avoid light). Taxic responses are restricted to organisms capable of movement.*

tropism A tropism is a growth or movement in **plants** that occurs due to a specific stimulus. *Positive tropism occurs towards the stimulus. Negative tropism occurs away from the stimulus.*

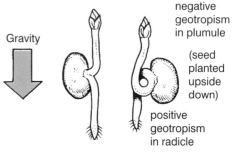

TROPISMS OF PLANTS

auxins are plant **hormones** which promote and regulate growth.

phototropism (or **heliotropism**) is the growth response of a plant to light. *Most leaves and stems (and some flowers like sunflowers) show this by bending and growing towards the light. It is thought that more **auxin** gathers on the side of the stem furthest from the light. Therefore that side grows faster, causing the stem to curve over towards the light.*

geotropism is the growth response of a plant to the pull of gravity. *The root of a plant is positively geotropic as it grows towards the pull of gravity. The stem is negatively geotropic as it grows away from the pull of gravity.*

hydrotropism is the growth response of a plant towards water. *Roots often grow sideways as there is more water closer to the surface.*

thigmotropism is the growth response of a plant to touch or contact. *For example, the **tendrils** of a pea plant curl around stems and other supports.*

clinostat A clinostat is a turntable device which rotates seedlings so that the effects of a stimulus are cancelled out. *It is used to study **tropisms**.*

tendril A tendril is a slender extension of a stem or leaf found in many climbing plants which shows **thigmotropism**.

insectivorous plants (or **carnivorous plants**) are plants that supplement their supply of nitrates by trapping and digesting insects. *Venus' fly trap has spiny hinged leaves that are **thigmotropic** and snap shut on alighting* *insects. The sundew plant has sticky hairs which curl around and trap an insect when it touches the hair.*

sense organs (or **sensory organs**) allow animals to detect their environment. *As well as **eyes**, **ears**, and **noses**, different animals have a variety of sense organs:*

- **Lateral line** is a series of receptor cells along the sides of fish and some amphibians. It is sensitive to pressure changes and so detects sound and water currents.
- **Palps** are jointed sensory organs found by the mouth parts of **arthropods** which act as smell or taste receptors.
- **Antennae** (singular: **antenna**) are long thin sense organs found on the heads of **insects** and many other **arthropods**. They are sensitive to touch, temperature, and chemicals.
- **Tentacles** are long flexible body parts in many invertebrate animals, especially **coelenterates** and **molluscs**. They are often used to manipulate objects but are also sensitive to touch.
- **Whiskers** (or **vibrissae**) are stiff hairs sensitive to touch, found on the noses of many animals, e.g. cats and burrowing animals.

nose The nose is the protuberance on the face that contains the **olfactory organ** (organ of **smell**). *The nose opens into a nasal cavity lined with a mucous membrane which helps trap dust, germs, etc.*

olfactory organ is the organ of **smell** situated in the roof of the nasal cavity. *Extending from the roof are olfactory hairs containing the **dendrites** of sensory neurones. These act as chemoreceptors sensitive to volatile substances. Nervous impulses from the olfactory nerve are interpreted by the brain as sensations of smell. The human sense of smell is much more extensive than that of taste. Many flavours in food have as much to do with the sense of smell as with the sense of taste.*

Human sense	Sensory organ
sight	**eye**
hearing	**ear**
touch	**skin**
taste	**tongue**
smell	**nose**

tongue The tongue is the muscular organ of taste which is attached to the floor of the mouth. *It is also used in manipulating food during chewing and swallowing, and in humans is involved in speech. The surface of the tongue is covered with taste receptors called **taste buds**.*

taste bud A taste bud is a small sense organ on the tongue. *It contains chemoreceptors which are sensitive to four types of taste: sweet, sour, salt, and bitter. Taste buds for each taste are concentrated in certain areas of the tongue. Nervous impulses from the taste buds are interpreted by the brain as taste sensations.*

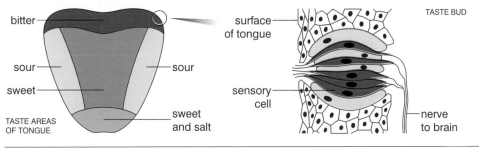

bitter — sour — sweet — surface of tongue — sensory cell — sour — sweet and salt — TASTE BUD — nerve to brain

TASTE AREAS OF TONGUE

silicon (symbol: **Si**) is the second element in group IV of the periodic table and is the second most abundant element in the Earth's crust (nearly 28%). *Silicon is mainly found combined with oxygen in the form of silicon dioxide in **quartz** and **sand**. It also occurs in the form of **silicates** in many **rocks** and **clays**. Most compounds of silicon are giant 3D structures. The element silicon is often described as a **metalloid**, as its properties are intermediate between metals and non-metals. It can exist as a hard, shiny solid which conducts electricity. Most of its compounds are **covalent**.*

silicon chip (or **microchip**) A silicon chip is a single crystal of very pure semiconducting silicon which has miniature electronic circuits printed on to it. *Silicon chips are extensively used in most **electronic systems**. They are made by slowly cooling very pure molten silicon. The first crystals to form during this recrystallisation process are very pure and are therefore ideal for use as silicon chips.*

quartz is the natural crystalline form of **silicon dioxide** (**silica**). *It is one of the hardest of common minerals. Quartz is a **macromolecule** in which each silicon atom is covalently bonded to four oxygen atoms. This tetrahedral arrangement is like that of diamond, and gives quartz a transparent crystalline structure with a high melting point.*

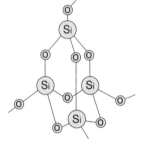

STRUCTURE OF SILICON DIOXIDE (QUARTZ)

sand is an impure form of quartz derived from the weathering of quartz-bearing rock. *Chemically it is mainly **silicon dioxide**. Most particles of sand have diameters in the range 0.06–2.00 mm. Sand is often yellow or red in colour due to iron(III) oxide. The element **silicon** can be extracted from a pure form of sand called 'white sand' by **reduction** with magnesium. Sand is used in mortar, concrete, and in making **glass**.*

silicates are compounds of the elements silicon, oxygen, and a metal. *Natural silicates form the main components of most **clays** and rocks. All silicate minerals are based on one fundamental structural unit – the SiO_4 tetrahedron. Silicate minerals are classified according to how the tetrahedra are linked together and which metal ions are interspersed.*

clay is a fine-grained sedimentary deposit which has **silicate** chains interspersed with aluminium ions (aluminosilicate). *Most clays are brown in colour due to the presence of iron.*

kaolin (or **china clay**) is a soft, white form of **clay** composed mainly of kaolinite (hydrated aluminium silicate). *It is widely used for making china and porcelain, and in some medicines.*

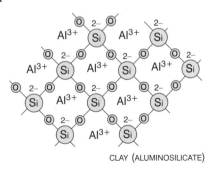

CLAY (ALUMINOSILICATE)

asbestos is a fibrous **silicate** which has its silicate chains interspersed with calcium and magnesium ions (see diagram). *Asbestos is chemically inert and fire-resistant and was formerly used in fire-protective clothing and brake linings. However, asbestos dust contains microscopic fibres that can irritate the lungs and cause the lung disease asbestosis, so its use is now restricted.*

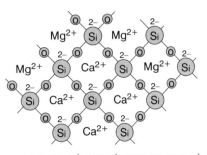

ASBESTOS (CALCIUM/MAGNESIUM SILICATE)

glass is a transparent or translucent material consisting mainly of calcium and sodium **silicates**, and is a non-crystalline solid in which the atoms are arranged randomly. *Glass is sometimes referred to as a 'supercooled liquid' because it has no well-defined melting point. It is made by heating together **sand** (silicon dioxide), limestone (calcium carbonate), and soda (sodium carbonate) at 1500°C.*

limestone	+	sand	→	calcium silicate	+	carbon dioxide
$CaCO_3$	+	SiO_2	→	$CaSiO_3$	+	CO_2
soda	+	sand	→	sodium silicate	+	carbon dioxide
Na_2CO_3	+	SiO_2	→	Na_2SiO_3	+	CO_2

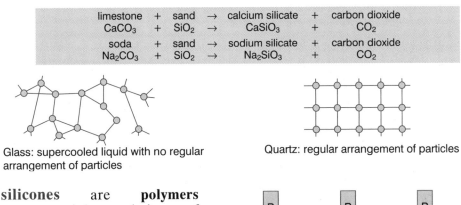

Glass: supercooled liquid with no regular arrangement of particles

Quartz: regular arrangement of particles

silicones are **polymers** containing chains of silicon atoms alternating with oxygen atoms, with organic alkyl groups linked to the silicon atoms. *Silicone polymers are very water-repellent (like silicon grease) and fire-resistant (unlike normal carbon polymers they produce no toxic fumes). Smaller silicone molecules are clear oily liquids and are used in cosmetic creams and paints. Larger molecules form rubbery solids or waxes, some of which are used by surgeons for replacement skin and breast enlargement (silicone implants).*

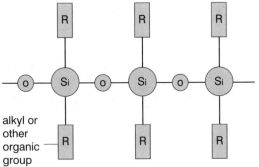

alkyl or other organic group

SILICONE POLYMERS

skeleton A skeleton is a structure in an animal that provides support for the body, protection for internal organs, and a framework for anchoring muscles and ligaments.

hydrostatic skeleton A hydrostatic skeleton is one in which support is provided by liquid under pressure in the body cavity. *The skeletons of soft-bodied land animals like slugs, earthworms, and caterpillars are of this type.*

exoskeleton An exoskeleton is an skeleton found on the outside of the animal. *Examples are the bony plates of tortoises, the shells of **molluscs**, the hard shell of **crustaceans**, and the hard **cuticle** of **arthropods**.*

cuticle The cuticle is a protective layer of hard material which covers arthropods. *It is made of chitin, a light but strong material, and forms the **exoskeleton**.*

ecdysis (or **moulting**) is the periodic loss of the cuticle in arthropods. *Animals with exoskeletons can only grow by shedding their outer cuticle. The animal grows while the new cuticle is still soft. It is then hardened.*

endoskeleton An endoskeleton is a skeleton that lies entirely inside the body of an animal. *The bony skeletons of vertebrate mammals (like the **human skeleton**) are endoskeletons.*

connective tissue is strong, tough tissue that holds organs or other tissues together. ***Bone** is a very hard connective tissue; **cartilage** is a less hard one, but is still tough.*

cartilage (or **gristle**) is a tough **connective tissue** which is softer than bone as it does not contain as many mineral salts. *Cartilage is found on the ends of bones where they meet in a joint. In some joints it is the main cushion between the bones. Cartilage also maintains the shape of certain organs like the nose or the ear flap (pinna), and forms the whole **endoskeleton** in some animals such as sharks.*

ossification is the gradual replacement of cartilage by bone. *In the skelton of embryos and young animals and children, the percentage of cartilage is much greater. Some of it is slowly turned to bone by the build-up of minerals.*

osteoporosis is a condition caused by low levels of calcium in the bones. *It makes the bones soft, weak, and fragile, and can be very painful. It is especially common in older women.*

ligament A ligament is a tough elastic structure of **connective tissue** that connects bones together at movable joints.

collagen is a **fibrous protein** found in connective tissue. *Collagen accounts for over 30% of the total body protein in mammals and is found in **skin**, **ligament**, **tendon**, **cartilage**, and **bone**.*

muscle is special elastic tissue that contracts or relaxes to produce movement. *This movement is stimulated by nervous impulses. In humans there are about 600 muscles which make up 40% of bodyweight.*

voluntary muscles (or **striated muscles**) are those whose action is controlled by conscious activity, such as the movement of the arms and legs.

involuntary muscles (or **smooth muscles**) are those whose action is not controlled by conscious activity, such as the movements of internal organs (e.g. stomach and intestines).

cardiac muscle is a special type of **involuntary muscle** found in the heart. *It can produce its own electrical impulses to provide constant rhythmical contractions without becoming tired.*

flexor muscles are muscles which flex (bend) a limb or part.

extensor muscles are muscles which extend (straighten) a limb or part.

antagonistic pairs are muscles that work in pairs but have an opposite effect to each other, like the **biceps** and **triceps**. *As one contracts the other relaxes.*

biceps is the **flexor muscle** in the upper arm which bends the forearm.

triceps is the **extensor muscle** in the upper arm which straightens the forearm.

tendon (or **sinew**) is tough **connective tissue** that connects a muscle to a bone. *Tendons consist of collagen fibres which are non-elastic, and therefore transmit the contraction or relaxation of the muscle to the bone.*

actin and **myosin** are two fibrous proteins found in muscle fibres. *They interlock to form the main part of the contraction mechanism.*

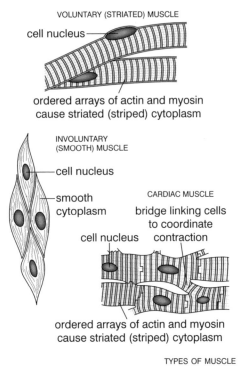

VOLUNTARY (STRIATED) MUSCLE

cell nucleus

ordered arrays of actin and myosin cause striated (striped) cytoplasm

INVOLUNTARY (SMOOTH) MUSCLE

cell nucleus

smooth cytoplasm

CARDIAC MUSCLE

bridge linking cells to coordinate contraction

cell nucleus

ordered arrays of actin and myosin cause striated (striped) cytoplasm

TYPES OF MUSCLE

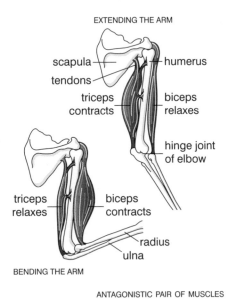

EXTENDING THE ARM

scapula

humerus

tendons

triceps contracts

biceps relaxes

hinge joint of elbow

triceps relaxes

biceps contracts

radius

ulna

BENDING THE ARM

ANTAGONISTIC PAIR OF MUSCLES

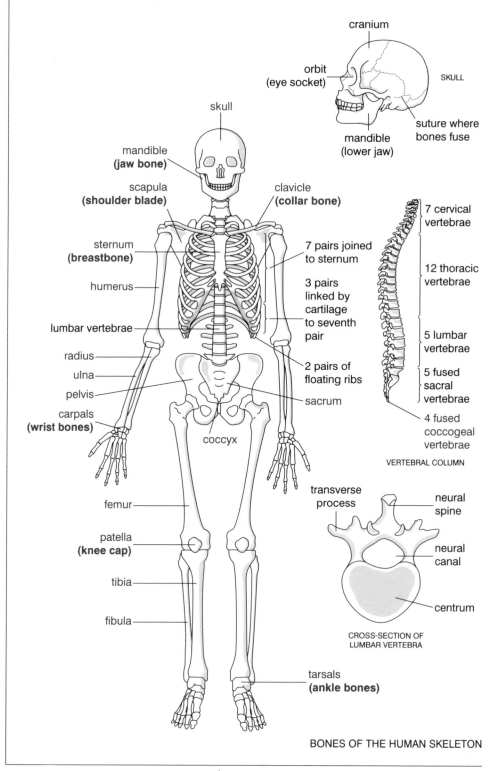

cranium

orbit
(eye socket)

SKULL

skull

mandible
(jaw bone)

mandible
(lower jaw)

suture where
bones fuse

scapula
(shoulder blade)

clavicle
(collar bone)

sternum
(breastbone)

7 pairs joined
to sternum

humerus

3 pairs
linked by
cartilage
to seventh
pair

lumbar vertebrae

radius

ulna

2 pairs of
floating ribs

pelvis

carpals
(wrist bones)

sacrum

coccyx

7 cervical
vertebrae

12 thoracic
vertebrae

5 lumbar
vertebrae

5 fused
sacral
vertebrae

4 fused
coccogeal
vertebrae

VERTEBRAL COLUMN

femur

transverse
process

neural
spine

patella
(knee cap)

neural
canal

tibia

fibula

centrum

CROSS-SECTION OF
LUMBAR VERTEBRA

tarsals
(ankle bones)

BONES OF THE HUMAN SKELETON

human skeleton The human skeleton is an **endoskeleton** consisting of 206 bones, the largest being the femur (thigh bone) and the smallest being the stapes in the middle ear. *Its main functions are support and protection of the body, movement, and the making of blood cells in the marrow of the long bones.*
- **axial skeleton** is the main longitudinal section of the skeleton, i.e. **skull**, **vertebral column**, and **rib cage**.
- **appendicular skeleton** is the term for those parts of the skeleton that are attached to the axial skeleton, i.e. **pectoral girdle**, **pelvic girdle**, arms and legs.

skull The skull is the skeleton of the head, consisting of the **cranium** and bones of the face and jaw. *Most of the bones in the skull are fused together at immovable joints called sutures. The only bone that can move is the **mandible** or lower jaw, which is fixed by a **hinge joint** to the rest of the skull.*

cranium (or **brain case**) The cranium is the part of the skull that encloses and protects the brain, consisting of eight fused bones.

vertebral column (or **backbone** or **spine** or **spinal column**) The vertebral column is a flexible series of small bones called **vertebrae**. *It supports the skull, protects the spinal cord, and provides points of attachment for the **pelvic** and **pectoral** girdles.*

vertebrae (singular: **vertebra**) are the bones that make up the vertebral column. *They generally have a thick body (centrum), a canal for the spinal cord, and projections for attachment of muscles. There are 33 in the human backbone.*

intervertebral discs of **cartilage** separate the vertebrae in the vertebral column. *They absorb shock and give flexibility. If one is displaced ('slipped disc') it causes severe pain.*

thorax The thorax is the chest region between the skull and the stomach which contains the heart and the lungs. *It is protected by the **rib cage**.*

rib cage The human rib cage is made up of 12 pairs of ribs, forming the walls of the thorax or chest area and protecting the heart, lungs, etc. *All the ribs are joined at the back to the thoracic **vertebrae**. The first seven pairs are also joined at the front to the **sternum** or breastbone, and the next three linked to it by cartilage. The last are small 'floating ribs'. Movement of the rib cage in breathing is controlled by **intercostal muscles**.*

pectoral girdle (or **shoulder girdle**) The pectoral girdle is the set of bones at the front end of a vertebrate's body which support the forelimbs. *In humans it consists of two dorsal **scapulae** (shoulder blades) attached to the spine and two ventral **clavicles** (collar bones) attached to the **sternum** (breastbone). Attached to the pectoral girdle are the arms (forelimbs).*

pelvic girdle (or **hip girdle**) The pelvic girdle is the set of bones at the rear end of a vertebrate's body which supports the hindlimbs.

pelvis The human pelvis forms the **pelvic girdle** and is joined to the base of the spine. *It is made up of three fused bones (ilium, pubis, and ischium), and the legs are attached to it.*

skeleton (bone and teeth)

bone is the hard **connective tissue** of which the skeleton of most vertebrates is made. *It contains **collagen** fibres and calcium salts.*

spongy bone is bone that has spaces in it, which may contain **bone marrow**. *Spongy bone is light and is found in short or flat bones like the breast bone or **sternum**. It also fills the ends of large bones like the thigh bone or femur.*

compact bone is bone which has few spaces in it and is found in the outer layer of most bones. *It is made up of concentric rings of tightly packed bone cells.*

bone marrow is soft tissue that fills the centre of **spongy bone**. *Bone marrow has a good supply of blood and makes red and white blood cells and blood platelets.*

joint A joint is the point of contact between two or more bones together with the tissue that surrounds it. *Some joints between bones are immovable joints, e.g. the sutures of the **skull**. Most joints in the body are movable joints. The bones in such joints are held together by **ligaments** and lubricated by the fluid from the **synovial membrane**.*

synovial membrane The synovial membrane is the lining of the sac which surrounds a movable joint. *It secretes the **synovial fluid**.*

synovial fluid is a liquid secreted by the **synovial membrane** which lubricates the joint and reduces friction between the bones when they are moving.

ball-and-socket joints are the most flexible of movable joints which allow bones to swivel and move in many directions. *The round head of one bone fits into a cup-shaped socket of another bone (see diagram). Examples are the hip joint and shoulder joint.*

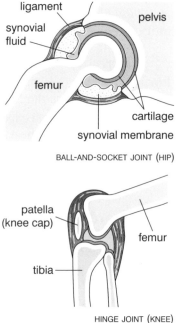

BALL-AND-SOCKET JOINT (HIP)

HINGE JOINT (KNEE)

hinge joints are movable joints which allow movement in a single plane only (like the hinge of a door). *Examples are the knuckle, knee, and elbow joints.*

gliding joints (or **sliding joints** or **plane joints**) are movable joints which allows flat bones to glide over one another. *Gliding joints occur between the **carpals** in the hand and the **tarsals** in the foot.*

tooth A tooth is a hard bone-like structure in a vertebrate which is mainly used for biting, chewing, and grinding food.

deciduous teeth (or **milk teeth**) are the first of two sets of teeth in a mammal. *In a child there are 20 milk teeth. At around 7 years of age these teeth begin to fall out and are replaced by **permanent teeth**.*

permanent teeth are the second and final set of teeth in a mammal. *An adult human normally has 32 permanent teeth consisting of incisors, canines, premolars, and molars.*

incisors are sharp chisel-shaped teeth at the front of the mouth used for biting and cutting.

canines (or **cuspids** or **eye-teeth**) are sharp cone-shaped teeth used to tear food. *Carnivores, like dogs, have highly developed canine teeth. Herbivores, like rabbits, do not have canine teeth.*

premolars (or **bicuspids**) are blunt broad ridged teeth in front of the molars, used for grinding and chewing food.

molars (or **tricuspids**) are blunt broad ridged teeth (like premolars) but with a larger surface area and at the rear of the mouth.

wisdom teeth are the four **molars** which appear last of all, sometimes into middle age, hence their name.

plaque is a sticky film of food debris, saliva and bacteria which forms on teeth after meals. *If not removed by brushing it hardens to form **tartar**.*

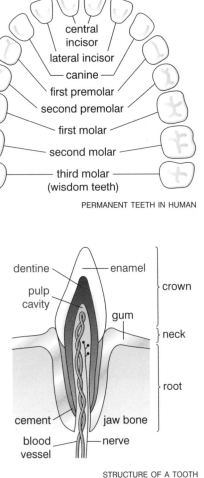

PERMANENT TEETH IN HUMAN

STRUCTURE OF A TOOTH

tartar is a hard deposit formed on the teeth and not removed by brushing.

tooth decay (or **dental caries**) is caused by bacteria in plaque. *If you eat sweet foods the **plaque** absorbs the sugar like a sponge. **Bacteria** in the plaque then feed on this sugar and convert it to acid. This dissolves away the protective **enamel** layer of the tooth, making a hole. If untreated, the acid gradually works its way into the **dentine** and **pulp cavity** of the tooth, causing infection and the death of the tooth.*

gum disease results if plaque is not removed and the bacteria present infect the gums. *The gums swell and when you brush your teeth they may bleed. If the **bacteria** are allowed to spread they could infect the root. The tooth becomes loose and may have to be extracted. Gum disease can be prevented by regular brushing of the teeth (two or three times a day), especially just after meals. Regular visits to the dentist allow treatment of any early signs of tooth decay or gum disease.*

skin The skin is the organ which makes up the outermost layer of the body. *It is the largest human organ which in an adult covers about 2 sq m. The skin contains many different structures and has a variety of important functions.*

FUNCTIONS OF THE SKIN

Protection	Outermost layer is tough and waterproof. It protects the body from injury, water loss and infection.
Sensitivity	Nerve endings in the skin can detect pain, temperature and pressure.
Homeostasis (water and temperature)	Sweat glands control water loss and help maintain body temperature.
Storage	Fat deposits are stored in the fatty tissue under the skin.
Vitamin production	Vitamin D is produced from the action of sunlight on a cholesterol derivative found in the skin.

structure of the skin Human skin is made of two main layers called the **epidermis** and **dermis**.

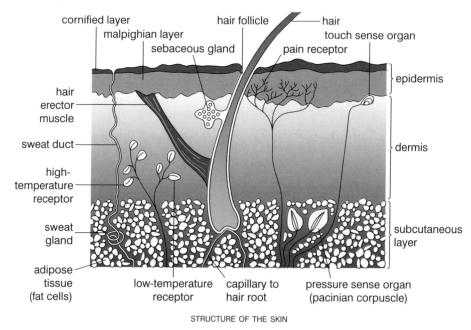

STRUCTURE OF THE SKIN

malpighian layer The malpighian layer is the uppermost living layer of the skin which produces new skin cells. *These new cells gradually move towards the surface of the skin. As they move upwards they die (average life about four weeks) and fill with a hard protein called keratin. This tough dead layer of cells is the **cornified layer**.*

cornified layer The cornified layer is the tough, outermost layer of skin, consisting of dead cells. *It is waterproof and protects the soft malpighian layer underneath. On those parts of the body that get the most wear, like the soles of the feet, the cornified layer grows very thick.*

hair follicle A hair follicle is a long narrow tube in the epidermis which contains a hair. *The hair grows as new cells are added at its base from cells lining the follicle.*

sebaceous gland A gland alongside the **hair follicle** which secretes an oily liquid (sebum) to keep the hair and skin soft and supple.

epidermis The epidermis is the outer protective layer of the skin. *It consists of a living layer of epithelial tissue (the **malpighian layer**) and a hard layer of dead cells (the **cornified layer**).*

dermis The dermis is a thick layer of **connective tissue** underneath the epidermis of the skin. *It contains elastic fibres. As a person ages these fibres lose their elasticity and the skin becomes wrinkled. The dermis contains blood capillaries and nerve endings.*

subcutaneous layer (or **adipose tissue**) The subcutaneous layer is the layer of fatty tissue underneath the skin. *It contains fat deposits which act as an insulating layer and helps keep body heat in. It is also a store of food.*

sweat gland A sweat gland is a small **exocrine gland** in the **subcutaneous layer** which secretes sweat. *Sweat is a watery fluid with small amounts of salt and urea. Sweat passes along a narrow tube (sweat duct) to the surface of the skin. It cools the body by evaporation.*

sense organs in the skin There are various nerve endings to different sense organs in all layers of the skin. *In the **epidermis** are pain receptors, in the **dermis** are high and low temperature receptors, and in the **subcutaneous layer** are pressure receptors.*

vasodilation is the opening of the **capillaries** in the dermis layer so that more blood flows and more heat is lost through the skin. *Vasodilation results in a decrease in blood pressure.*

vasoconstriction is the closing of the dermal **capillaries** so that less blood flows and less heat is lost through the skin. *This occurs when the body is cold. Vasoconstriction results in an increase in blood pressure.*

hypothermia is the gradual lowering of body temperature due to heat loss in cold weather. *Normal body temperature is 37°C, but if this drops by 2°C the person becomes drowsy due to a decrease in **metabolic rate**. If temperature drops further, the person becomes unconscious and may die. Warm drinks, many layers of clothes, and lots of activity prevent hypothermia.*

heat stroke is a failure of the body's temperature regulation in high temperatures. *The sweat glands fail to produce sweat and the person must be moved to cool surroundings.*

skin cancer can be caused by excessive ultraviolet radiation from the Sun which can damage the DNA in skin cells.

melanin is a brown pigment found in the cells of the **epidermis**. *Melanin absorbs harmful ultraviolet rays from the sun. People with fair skin produce more in direct sunlight, producing a suntan. People from tropical areas have melanin in all layers of the epidermis giving them dark skin.*

solar system The solar system is our **Sun** and the nine major planets that orbit around it: **Mercury**, **Venus**, **Earth**, **Mars**, **Jupiter**, **Saturn**, **Uranus, Neptune** and **Pluto**. *The solar system also includes **moons** of the planets as well as **asteroids** and **comets**. The Sun accounts for over 99% of the mass of this system, and Jupiter for more than half of the rest.*

Sun The Sun is a **star** at the centre of our solar system. *It is about 149 600 000 km from Earth and its diameter is about 110 times that of Earth. The temperature inside is about 15 million°C and on the surface is about 6000°C. The core is so hot that atoms are broken down into ions and electrons (**plasma**). The Sun is about 75% hydrogen and 25% helium, with less than 1% heavier elements. Light from the Sun takes around 8 minutes to reach Earth. Like all stars, the Sun emits a wide spectrum of **electromagnetic radiation**, not just visible light.*

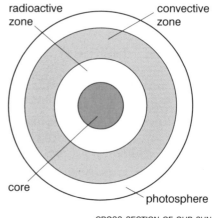

CROSS-SECTION OF OUR SUN

solar flares A solar flare is a burst of hot material that shoots out from the **Sun** into space. *Some solar flares are 400 000 km in length. Solar flares affect radio communication on Earth.*

sunspot A sunspot is a dark patch on the Sun's surface, resulting from a localised fall in temperature to about 4000°C. *Sunspots tend to occur in clusters and last about two weeks. The number of sunspots changes over an eleven-year cycle.*

solar wind The solar wind consists of charged particles that stream out from the **Sun** all the time but whose intensity varies with the month or time of year. *As the solar wind sweeps past the Earth it distorts magnetic fields and causes radio interference. It also causes the **aurora**.*

cosmic rays (or **cosmic radiation**) are high-energy particles that fall on the Earth from space. *Most of them are charged particles that have come from the Sun. More cosmic rays bombard the Earth from the west, due to the deflection of the particles in the Earth's magnetic field.*

aurora Auroras are luminous phenomena that occur in the night sky especially around the poles. *They result from the interaction of the **solar wind** with atoms and molecules high in the upper atmosphere, which are then attracted to the poles by the Earth's magnetic field.*

aurora borealis (or **northern lights**) is the aurora which occurs above the North Pole.

aurora australis (or **southern lights**) is the aurora that occurs above the South Pole.

planet A planet is a major celestial body that orbits the Sun in a slightly elliptical orbit. *There is an attractive force of **gravity** between the Sun and the planet, which acts in a direction towards the centre of gravity of the Sun. Planets orbit the Sun in the same plane (except Pluto) and reflect the Sun's light, but do not give out any light themselves. The four inner planets (Mercury, Venus, Earth, Mars) are relatively small, rocky and volcanic. Scientists call them 'rocky dwarfs'. The outer planets are giant balls of gas (except Pluto). Scientists sometimes call them 'gassy giants'. The further a planet is from the Sun, the longer is the period of its orbit (its 'year').*

Planet	Diameter (relative to Earth's diameter)	Distance from Sun (relative to Earth's distance)	Length of year (in Earth years)	Length of day (in Earth days)	Force of gravity (compared with Earth)	Number of moons	Average surface temperature (°C)	Atmosphere (main gases)
Mercury	0.39	0.39	0.24	59	0.38	0	350	none
Venus	0.97	072	0.61	243	0.86	0	460	thick carbon dioxide clouds
Earth	1	1	1	1	1	1	20	nitrogen and oxygen
Mars	0.53	1.52	1.88	1.03	0.38	2	−40	thin carbon dioxide
Jupiter	11.2	5.2	11.86	0.41	2.5	16 + 1 ring	−120	
Saturn	9.5	9.53	29.5	0.44	1.13	18 + 7 rings	−180	hydrogen
Uranus	3.7	19.2	84	0.67	1.04	15 + 11 rings	−210	helium and ammonia
Neptune	3.5	30	165	unknown	1.4	8 + 4 rings	−220	methane
Pluto	0.4	39	249	unknown	unknown	1	−230	none (frozen)

PLANETS OF OUR SOLAR SYSTEM

asteroid (or **minor planet**) The asteroids are a large number of rocks orbiting the Sun in a belt between the orbits of Mars and Jupiter. *They vary in size from between a few km to about 200 km.*

comet A comet is a small lump of ice and rock orbiting the Sun in a non-circular orbit and in a different plane from that of the planets. *As a comet nears the inner part of our solar system, the Sun heats it up to produce a tail of gas. As the comet moves in space, its tail streams away from the Sun. Comets travel faster when closer to the Sun as the pull of gravity speeds them up. Halley's comet can be seen from Earth every 76 years as it passes close to Earth (last time in 1986).*

meteor (or **shooting star**) A meteor is a particle from space which enters the Earth's atmosphere and becomes so hot from friction with air particles that it glows white. *It appears as a streak of light in the sky. Most meteors burn up before they reach the Earth's surface.*

meteorite A meteorite is a rock formed when a large **meteor** does not burn up completely in the atmosphere and reaches the Earth's surface.

sound and ultrasound

sound is a **progressive longitudinal wave** caused by the vibration of an elastic **medium** such as air.

sound waves consist of **compressions** and **rarefactions** caused in a medium when it is disturbed by a vibrating object. *They are normally drawn as a series of wavefronts which mark the regions of compression of the wave. The distance between each wavefront is the **wavelength** of the sound.*

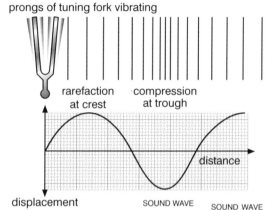

compression is the squashing together of particles (e.g. those in the medium of a **longitudinal wave**).

rarefaction is the spreading apart of particles (e.g. those in the medium of a **longitudinal wave**).

speed of sound The speed of sound depends upon the **density** and temperature of the medium. *Sound waves cannot travel in a vacuum because there is no*

Medium	Density (g cm⁻³)	Speed of sound (at 0°C)
air	0.001	330 m s⁻¹
water	1.0	1400 m s⁻¹
brick	5.6	3700 m s⁻¹
concrete	7.8	5000 m s⁻¹

material (medium) to vibrate. In general, sound travels fastest in solids, slower in liquids, and slowest in gases. This is because the particles in solids are closest together so the vibrations of the sound are passed on more rapidly. The speed of sound in air increases with temperature but is unaffected by pressure. Sound travels about a million times slower than light, which is why you see lightning before you hear thunder.

echo An echo is a **reflection** of a sound wave by a surface or an object so that a weaker version is detected after the original.

echolocation is a method of using the reflection of waves off an object to determine its exact position. *If you measure the distance there and back to the object and the time taken for an echo to return, the speed of sound is the distance divided by the time. Bats use echolocation when flying and hunting at night.*

subsonic describes a speed below the speed of sound in the same medium.

supersonic describes a speed greater than the speed of sound in the same medium.

sonic boom A sonic boom is the loud bang caused by the shock wave produced by an aircraft travelling at supersonic speeds. *At **supersonic** speeds the aircraft overtakes its own sound waves. This causes a build-up of pressure on the front of the aircraft which is unable to escape. This is the shock wave which a listener hears as a sudden loud sonic boom. The* *loudness depends on the speed and altitude. Aircraft are not permitted to go 'supersonic' over land.*

Doppler effect The Doppler effect is the change in **frequency** of a sound caused by either the listener or the source moving relative to the other. *As a police car moves towards you, the frequency of the sound increases (the sound becomes higher). As the car moves away from you the frequency of the sound decreases (the sound becomes lower). The effect is named after the Austrian physicist Christian Doppler (1803–53).*

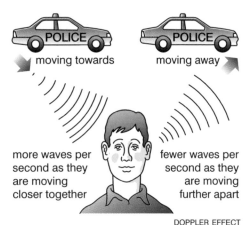

moving towards

moving away

more waves per second as they are moving closer together

fewer waves per second as they are moving further apart

DOPPLER EFFECT

infrasound is sound below the threshold of the **human hearing range**, around 20 Hz. *Infrasound is used by whales for communication. Such waves have few uses as they are distressing to humans.*

ultrasound is sound above the **human hearing range**, around 20 000 Hz. *Such sounds can be detected by animals such as bats and dogs. Ultrasound travels as wave with frequencies between 20 kHz and 10 MHz. Such waves have a wide range of uses.*

- **ultrasound scanning** is used during pregnancy to check the development of unborn babies. It is painless, and safer than X-rays. The detector can be tuned to pick up only frequencies reflected from one particular source such as bone or fat or muscle, which reflect ultrasonic waves at particular speeds. It can therefore distinguish between these tissues. The reflected signals are processed by a computer which produces an image (scan) of the baby inside the womb.

- **ultrasonic cleaning** uses the high-frequency vibrations of ultrasound to shake the dirt from clothing or other materials, and may be used by dentists to clean the coating of tartar off your teeth.

- **ultrasonic stress detection** uses ultrasound to detect internal cracks in metal parts such as aircraft bodies and wings. These cracks are stress fractures caused by continuous flexing.

sonar (abbreviation for **SO**und **NA**vigation and **R**anging) is a system which uses **ultrasound** for **echolocation** to detect underwater objects or to determine the depth of the water.

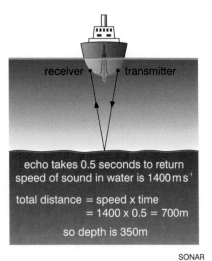

receiver   transmitter

echo takes 0.5 seconds to return
speed of sound in water is 1400 m s⁻¹

total distance = speed x time
            = 1400 x 0.5 = 700m

so depth is 350m

SONAR

sound and musical notes

loudness is a property of a sound wave determined by its amplitude. *The greater the* ***amplitude***, *the louder the sound. The sensation of loudness in the ear is subjective and depends on the sensitivity of the ear to the particular frequencies. Loudness is often measured in* ***decibels***.

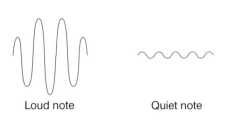

Loud note   Quiet note

pitch is a property of sound determined by its frequency. *High-pitched sounds are associated with high frequencies, low-pitched sounds (deep, low notes) with low frequencies.*

High pitched note   Low pitched note

SOUND CHARACTERISTICS

decibel (symbol: **dB**) A decibel is a commonly used unit of sound intensity or loudness. *Each 10 dB increase represents a 10-fold in the energy (loudness) of the sound.*

Sound (pitch)	Frequency
upper limit of hearing	20 000 Hz
whistle	10 000 Hz
high note (treble)	1000 Hz
low note (bass)	100 Hz
drum beat	20 Hz

beats are regular variations in loudness when two or more sounds of slightly different frequency are heard at the same time. *It is the result of* ***interference*** *between the two sound waves, as the waves go alternately in and out of phase with each other. The closer together the frequency of the two sounds, the slower the beats.*

Sound (loudness)	dB level
threshold of hearing	0 dB
bird singing	30 dB
normal conversation	60 dB
road drill	80 dB
loud thunderclap	110 dB
threshold of pain	130 dB

octave An octave is the interval between two **musical notes** such that the higher note has double the frequency of the lower note. *Two notes an octave apart blend very well to the human ear. Middle C has a frequency of 256 Hz, and the C above it has a frequency of 512 Hz (an octave higher).*

Two waves of two different frequencies

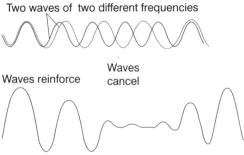

Waves reinforce   Waves cancel

Resultant wave form

BEATS

musical scale A musical scale is a series of notes arranged in order of increasing or decreasing frequency.

vibrating strings produce sound waves when stretched and made to vibrate (e.g. in a piano or guitar). *If the vibrating length of a string is halved, then the frequency of the note doubles and increases by one* ***octave***. *If the tension in the string is increased, the frequency increases, but if its thickness increases, then the frequency decreases.*

sonometer A sonometer is an apparatus for investigating stretched wires or strings as sources of sound.

node A node is a point on a **standing wave** where the amplitude is zero and there is no vibration.

antinode An antinode is a point on a **standing wave** where the amplitude of the vibration is a maximum.

harmonics (or **overtones**) are frequencies of a wave which are multiples of the fundamental frequency. *When a stretched spring vibrates the simplest vibration is the 'fundamental' or first harmonic.*

quality of sound (or **timbre**) The quality or 'timbre' of a sound is a result of the harmonics which are present. *Middle C played on a piano has a different sound to Middle C played on a guitar because of the different harmonics.*

reverberation is the persistence of a sound for a longer period than normal. *It occurs when the time taken for an **echo** to return back to the source is so short that the original and the reflected wave cannot be distinguished. Reverberation in halls is a nuisance as it makes speech and music indistinct. It can be reduced by using soft materials, such as curtains, carpets, thick cushions or seats, etc., to absorb the sound energy and minimise the echoes.*

natural frequency is the frequency at which an object will vibrate if a vibration is started by a brief external force.

resonance occurs when a system is made to vibrate at its **natural frequency** by vibrations from another source of the same frequency. *A glass can be made to resonate by the effect of sound waves such as a singer's voice at this frequency. When you flick a wine glass with your finger from the outside, it will vibrate and ring at its natural frequency. Resonance can occur in mechanical and electrical systems as well as material objects.*

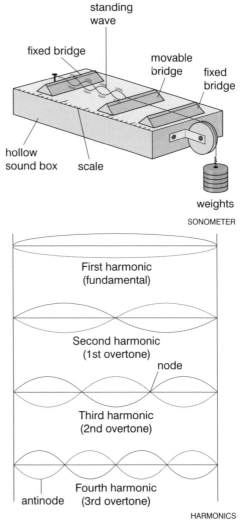

standing wave

fixed bridge

movable bridge   fixed bridge

hollow sound box   scale

weights

SONOMETER

First harmonic (fundamental)

Second harmonic (1st overtone)

node

Third harmonic (2nd overtone)

antinode   Fourth harmonic (3rd overtone)

HARMONICS

stars and the Universe

star A star is a celestial body that generates its own light and heat from **nuclear fusion** within its core. *Stars are not distributed uniformly throughout the universe but are collected together in **galaxies**.*

constellation A constellation is a group of **stars** in the sky which form a fixed pattern in relation to each other, as viewed from the Earth. *They may not actually be close together in space.*

protostar A protostar is a cloud of hot dust and gas (**gaseous nebula**) collected together under the force of gravity. *When it becomes dense enough, nuclear fusion begins and it becomes a star.*

stellar evolution (or **life of a star**) This consists of a series of changes that occur to a star during its lifetime, from birth to extinction. *There are two types depending on size of **protostar**.*

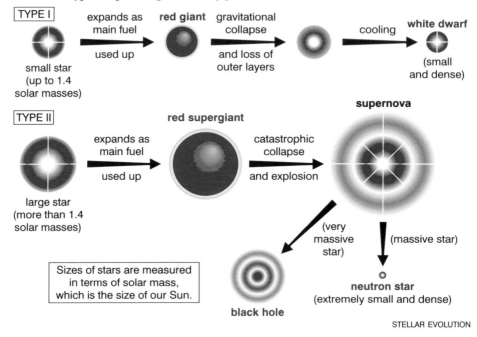

STELLAR EVOLUTION

Hertzsprung–Russell diagrams are graphs which plot the brightness of the stars against their colour. *Blue stars are very hot and at the beginning of their life. A red star is cooler and generally older. Our Sun is a yellow star in the middle part of its life (main sequence).*

galaxy A galaxy is a giant collection of gas, dust, and stars held together by gravitational attraction between its components. *Galaxies are not spread evenly throughout the universe but are grouped in clusters. There are over a billion galaxies in the Universe.*

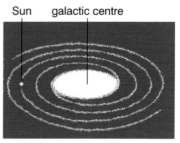

SPIRAL GALAXY

Milky Way The Milky Way is the **galaxy** to which our Sun belongs. *It is a spiral galaxy with a central disc and curving spiral arms. The Sun is in one of these arms. The main part of the galaxy is visible from Earth as a faint band across the night sky (the 'Milky Way'). Our galaxy is thought to contain around 10^{11} (100 billion) stars and to be around 100 000 light years across. It is part of a cluster of about 20 galaxies.*

light-year A light-year is an astronomical unit of distance (not time) and is equal to the distance travelled by light in one year. *The nearest star, called Proxima Centauri, is 4.2 light years from us. The nearest galaxy, called Andromeda, is about 1 million light years away.*

parsec A parsec is an astronomical unit of distance equal to 3.26 **light-years**.

Universe The Universe is all the matter, energy and space that exists. *It is estimated that our universe contains 10^{41}kg of mass collected into 10^9 (a billion) galaxies.*

outer space is the part of the **universe** that lies outside the Earth's atmosphere. *Between objects in space (planets and stars) there is very little matter, almost a **vacuum**.*

expansion of the Universe The Universe appears to be expanding because all the galaxies are getting further apart from each other.

red shift The red shift is a lengthening of the wavelength of light from distant stars so that it seems to shift towards the red end of the spectrum. *It is caused by a form of **Doppler effect** and demonstrates the **expansion of the Universe**.*

big-bang theory The big-bang theory suggests that the **universe** was formed from a highly dense central mass (the size of an atomic nucleus containing all the matter in the Universe) that exploded around 15 thousand million (15 billion) years ago.

steady-state theory The steady-state theory proposed that the **Universe** has always existed in a steady state with no beginning, and will have no end. *It suggests that matter is being created as the universe expands. It has lost favour to the **big-bang theory** as it has failed to account for the evidence of evolution in the Universe.*

black hole A black hole is a region of space where gravity is so strong that even light cannot escape. *It is thought to be formed after a **supernova**.*

supernova A supernova is an immense explosion which results when an old and very massive star uses up most of its fuel for nuclear fusion and collapses under the force of its own gravity.

radio astronomy is the study, using radio telescopes, of the radio waves emitted from celestial objects including **galaxies** and **quasars**.

quasar A quasar is an extremely bright and distant celestial object emitting radio waves. *Quasars were first discovered in 1961.*

pulsar A pulsar is a celestial object which emits regular pulses of radio waves (frequency 0.03 s to 4 s). *Pulsars were first discovered in 1968.*

macromolecules (or **giant structures**) are very large molecules often containing many thousands of atoms. *Natural and synthetic **polymers** are macromolecules, as are giant covalent molecules like **diamond** and **graphite**.*

crystals are pure solids with a regular lattice structure giving a regular polyhedral shape. *All crystals of a particular substance grow so that they have the same regular arrangement of molecules, ions, or atoms, and so have the same angles between their faces.*

dislocation A dislocation is an imperfection or discontinuity in a crystal lattice. *When molten substances solidify **crystals** form, but when they meet there is not a perfect fit, which results in a dislocation.*

diamond is a natural transparent form of the element carbon, each diamond crystal being one single **macromolecule**. *Inside such a molecule each carbon atom is tetrahedrally joined by four **single covalent bonds** to four other carbon atoms. Because it is so stable it is the hardest naturally occurring substance. Diamond is therefore used for cutting and grinding tools.*

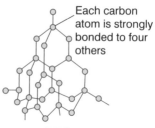

Each carbon atom is strongly bonded to four others

CARBON ATOMS IN DIAMOND

graphite (or **black lead** or **plumbago**) is a natural black form of the element carbon which exists as a **macromolecule** with a layered structure of flat hexagons. *The bonds inside each hexagon are strong **covalent bonds**. The bonds between the layers are weak **van der Waals' forces**, which makes graphite soft and flaky and suitable for use as a lubricant or in pencil leads. Graphite is also a good **conductor** of electricity and heat, as each carbon atom has only used three of its four outermost electrons in covalent bonding to three other carbon atoms. The spare electron can become delocalised along the flat hexagons, which therefore conduct electricity. Graphite when aligned along its hexagons can form very strong and light carbon fibres.*

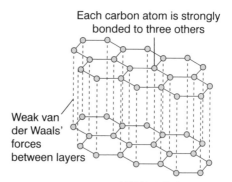

Each carbon atom is strongly bonded to three others

Weak van der Waals' forces between layers

CARBON ATOMS IN GRAPHITE

COMPARISON OF DIAMOND AND GRAPHITE

Property	Diamond	Graphite
appearance	transparent	black, shiny
hardness	extremely hard	very soft
density	$3.5\,g\,cm^{-3}$	$2.3\,g\,cm^{-3}$
conductivity	non-conductor	conductor
burning in oxygen	very difficult	easy, to form carbon dioxide

allotropes are solid forms of an element with different molecular structures. *Diamond and graphite are allotropes of carbon. Other allotropic elements are* **oxygen** *(normal dioxygen O_2 and ozone O_3),* **sulphur** *(rhombic and monoclinic forms), and* **phosphorus** *(red and white forms).*

fullerenes form a third **allotrope** of carbon consisting of spherical clusters of carbon atoms, which were discovered in 1985. *The most symmetrical is buckminsterfullerene which is composed of 60 carbon atoms bonded together. It is named after the American architect Buckminster Fuller (1895–1983), because it looks like the symmetrical 'geodesic' domes which he designed. It can be made by electrically evaporating graphite electrodes in helium gas at low pressure. Scientists believe that these fullerenes will have important uses as lubricants, catalysts, semiconductors, and superconductors.*

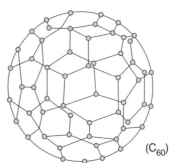

(C_{60})

Buckminsterfullerene: 60 carbon atoms covalently bonded to form 20 hexagons and 12 pentagons. They fit together to look like the surface of a football, hence the nickname **buckyballs**.

close packing is the packing of particles such as atoms or ions so as to occupy the minimum amount of space. *There are two main types, cubic close packing and hexagonal close packing, which are the structures adopted by most* **metals**.

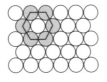

HEXAGONAL
CLOSE PACKING
(each particle is
surrounded
by six others)

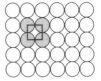

CUBIC
CLOSE PACKING
(less tightly packed
so can form a cube)

lattice A lattice is a structure in which there is a regular arrangement of atoms, ions, or molecules.

metallic lattice (or **atomic lattice**) A metallic lattice has atoms of the metal closely packed together in layers surrounded by a sea of electrons. *This* **close packing** *of atoms explains why many metals have a high density. There are also strong forces between these atoms which is why most metals have high melting points. These forces, although they are strong, are not rigid: when a force is applied, the atoms can slip over one another. This allows the metal to be malleable (made into sheets) and ductile (drawn into a thin wire) without cracking or breaking. The free electrons can readily conduct heat and electricity.*

ionic crystals are crystals composed of ions regularly arranged in a giant **lattice**. *Each ion is surrounded by oppositely charged ions. Ionic crystals have strong electrostatic forces between their ions (**ionic bonds**). They therefore usually have high melting and boiling points. Because their ions are in fixed positions, these crystals do not conduct electricity.*

studying science (accuracy and errors)

accuracy is the degree to which a measurement represents the actual value of the thing being measured. *Experimental measurement is always subject to some error. Some errors may be due to carelessness such as misreading a scale. Others may arise because of the inaccuracy of the measuring instrument itself. It is important to minimise reading errors and to appreciate the accuracy of the measuring instrument you are using.*

parallax error A parallax error is one which occurs when the eye is not placed directly opposite the scale from which a reading is being made.

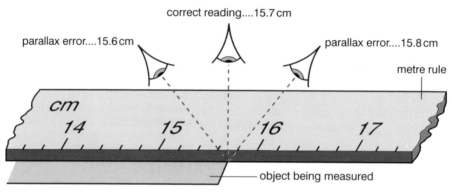

PARALLAX ERROR

To overcome this problem, instruments often have a mirror behind the pointer. The correct reading is obtained by making sure that the eye is exactly in front of the pointer, so that the reflection of the pointer in the mirror is hidden behind it. To avoid parallax errors when reading the height of a liquid, always read to the level of the centre of the meniscus (the curved level of liquid inside the tube).

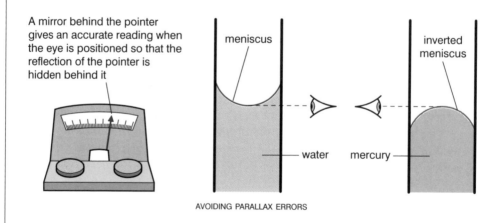

AVOIDING PARALLAX ERRORS

reading error A reading error is one due to the guesswork involved when a reading lies between two values on a marked scale.

Vernier caliper is closed and should read 0.0 mm. However, it reads 0.3 mm so the caliper has a zero error of + 0.3 mm

Vernier caliper reads 4.34 cm, but 0.3 mm must be subtracted from this reading to give a true reading of 4.31 cm

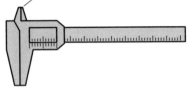 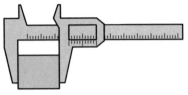

ZERO ERROR

zero error A zero error is one caused when a measuring instrument does not start from exactly zero. *This is shown for a **vernier caliper**.*

percentage error is a figure representing the difference between the value determined experimentally and the theoretical value.

$$\text{percentage error} = \frac{\text{difference between experimental and theoretical}}{\text{theoretical value}} \times 100$$

digital reading A digital reading is one produced by an electronic display of numerals. *A digital reading gives a precise figure, but it may not be exactly accurate. For example, an electronic balance which reads to 0.1 g will read the mass of a certain object as either 3.1 g or 3.2 g. Its scale is not continuous and it cannot distinguish between masses of less than 0.1 g. If the balance reads 3.1 g then the actual mass of the object may vary between 3.05 g and 3.14 g (to two decimal places).*

analogue reading An analogue reading is one which is given by an instrument which has a continuous scale. *Measuring cylinders, thermometers, rulers, clocks with hands, and meters with a scale and a pointer are all analogue instruments. They are subject to **reading errors**.*

significant figures (abbreviation: **sf**) The number of significant figures in a value is the number of digits which express its value to a particular level of accuracy, ignoring any leading or trailing zeros and disregarding the position of the decimal point. *For example, the value 0.0012 has two significant figures: the zeroes do not add to the accuracy but just indicate the size of the value.*

significant figures		decimal places
5	9.3925	4
	rounded to	
4	9.393	3
	rounded to	
3	9.39	2
	rounded to	
2	9.4	1
	rounded to	a whole
	9	number

decimal places (abbreviation: **dp**) The number of decimal places in a value is the number of digits after the decimal point.

rounding is the process of reducing the number of digits quoted in a particular reading. *We can round up to a certain number of decimal places or of significant figures. The last digit required is increased by one if the digit following it is 5 or more. Too many digits in a value may be misleading if the measurement is not really as accurate as that.*

scalar quantity A scalar quantity is one which has magnitude (size) but not direction. *Examples are **temperature, mass, density, energy,** etc.*

vector quantity A vector quantity is one which has both magnitude and direction. *Examples are **displacement, velocity, acceleration, force,** etc. When a value is given to a vector quantity, the direction must also be shown, normally by an arrow. The length of this arrow indicates the magnitude of the vector quantity.*

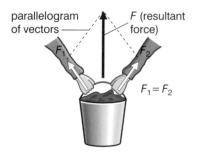

LIFTING A BUCKET WITH TWO HANDS

SI units (Système International d'Unités) are the international system of **units** recommended for all scientific work. *This system of units is derived from the m.k.s. units (metric, kilogram, second) and has now replaced the similar c.g.s. units (centimetre, gram, second) and the Imperial units. There are seven **base units**. There are also two supplementary units and 18 other derived units, each with its own special name. Each unit has its own symbol (usually one or two letters). Decimal multiples of the units are indicated by **prefixes**.*

Basic SI unit	Symbol	Quantity	Standard
metre	m	length	Distance light will travel in a vacuum in 1/299 792 458 of a second.
kilogram	kg	mass	Mass of international prototype kilogram made of platinum-iridium alloy and kept at Sèvres, France.
second	s	time	Time taken for 9 192 631 770 resonance vibrations of an atom of caesium–133.
kelvin	K	temperature	1/273.16 of the temperature of the triple point of water.
ampere	A	electric current	Current that produces a force of 2×10^{-7} newtons per square metre between parallel conductors which are one metre apart.
mole	mol	amount of substance	Amount of substance that contains the Avogadro number of particles (atoms, ions, or molecules).
candela	Cd	luminous intensity	Monochromatic source of light of frequency 540×10^{12} Hz with a power of 1/683 watt per steradian.

BASE UNITS

base unit A base unit of measurement is not defined in terms of other units but has its own standard definition. *For example, the metre is a base unit of length, but speed, which is measured in metres per second, is a derived unit. There are seven base units in the SI system (see table).*

Prefix	Symbol	Index notation	Meaning
tera-	T	10^{12}	1 000 000 000 000
giga-	G	10^9	1 000 000 000
mega-	M	10^6	1 000 000
kilo-	k	10^3	1 000
hecto-	h	10^2	100
deca-	da	10	10

Prefix	Symbol	Index notation	Meaning
deci-	d	10^{-1}	0.1
centi-	c	10^{-2}	0.01
milli-	m	10^{-3}	0.001
micro-	μ	10^{-6}	0.000 001
nano-	n	10^{-9}	0.000 000 001
pico-	p	10^{-12}	0.000 000 000 001

PREFIXES

prefix A prefix is a group of letters placed in front of a word to make a new word which has a different meaning. *Prefixes are used as multipliers to the base SI units as shown in the tables. The symbols for prefixes are small letters, except for the largest ones.*

conventions for units and symbols The names of units are not written with a capital letter even if they are named after people, but if a symbol is a capital letter it must always be so written. *A full stop is not used after a unit symbol (except at the end of a sentence) and there is no plural form of a unit symbol, e.g. 30 N or 40 kg. When writing numbers with SI units the digits are arranged in groups of three, and a space is placed between each group, e.g. 657 541.37 m.*

Physical quantity	Name of SI unit	Symbol	Named after
current	ampere	A	French physicist Andre Ampère (1775–1836)
radioactivity	becquerel	Bq	French physicist Henri Becquerel (1852–1908)
electric charge	coulomb	C	French physicist Charles de Coulomb (1736–1806)
capacitance	farad	F	English scientist Michael Faraday (1791–1867)
frequency	hertz	Hz	German physicist Heinrich Hertz (1857–96)
energy (work)	joule	J	English physicist James Prescott Joule (1818–89)
temperature	kelvin	K	English scientist Lord Kelvin (1824–1907)
force	newton	N	English scientist Sir Isaac Newton (1642–1727)
resistance	ohm	Ω	German physicist Georg Ohm (1787–1854)
pressure	pascal	Pa	French physicist Blaise Pascal (1623–62)
power	watt	W	English engineer James Watt (1736–1819)

UNITS NAMED AFTER PEOPLE

conversion factors to SI units Although in all science exams SI units will be used, it is sometimes useful to be able to convert other units into SI units.

From	To	Multiply by
inch	metre	2.54×10^{-2}
foot	metre	0.3048
square foot	square metre	9.2903×10^{-2}
litre	metre cubed	10^{-3}
gallon	litre	4.54609
miles/hr	metres/second	0.47704
km/hr	metre/second	0.27728

From	To	Multiply by
pound	kilogram	0.453592
$g\,cm^{-3}$	$kg\,m^{-3}$	10^3
horsepower	watt	745.7
mm Hg	pascal	133.322
atmosphere	pascal	1.01325×10^5
kW hour	joule	3.6×10^6
cal	joule	4.1868

studying science (measurement)

measurement of length depends upon the magnitude of the length. *A metre rule can measure to the nearest division, which is 1 mm. The **reading error** is therefore to the nearest 0.5 mm, which for smaller lengths is unacceptable. For more accurate readings of smaller lengths we use either a **vernier caliper** or a **micrometer screw gauge**.*

vernier caliper A vernier caliper is a measuring device which uses a **vernier scale** to measure to the nearest 0.1 mm. *This permits more accurate readings than a simple calibrated scale. It is a small movable scale graduated in intervals that are $\frac{9}{10}$ of those on the main scale. If this main scale reads to 1 mm, then the vernier reads to 0.1 mm.*

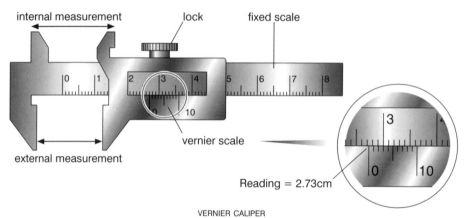

VERNIER CALIPER

micrometer screw gauge A micrometer screw gauge is an instrument with a movable circular vernier for accurate measurement of lengths of up to about 30 mm to the nearest 0.01 mm. *It can be used to measure accurately the thickness (diameter) of wires, etc.*

measurement of mass is achieved using a balance, which in a school laboratory may weigh to a milligram (0.001 g). *Common school balances are the beam balance (scale pans hung from a centrally pivoted bar) and electronic balances. Most electronic balances have a 'tare' by which the mass of the empty container can be stored in the balance's memory and automatically deducted from the mass of the container plus its contents.*

measurement of force is usually carried out using a spring balance or newton balance. *Such balances contain a helical spring whose extension is directly proportional to the force applied, provided that the spring is not overstretched (see **Hooke's law**).*

measurement of volume depends on the physical state of the substance (solid, liquid, or gas). *The volume of a substance is the amount of space it occupies and is normally measured in cubic centimetres (cm^3) or cubic metres (m^3). Volumes of gases can be measured using a gas syringe. Volumes of liquids can be measured approximately using a beaker, or more accurately using a measuring cylinder, **burette**, or **pipette**.*

microscope A microscope is an instrument which produces a magnified image of a small object which is often invisible to the naked eye.

compound microscope A compound microscope uses two lenses (objective and eyepiece). *The objective lens produces a real image which is then magnified by the eyepiece lens to produce a final much larger virtual image.*

magnification value The magnification value of a microscope tells you how many times bigger it makes the object. *To find the total magnification of the microscope you multiply the magnification of both the objective and eyepiece lenses. For example, if the eyepiece is 10×, objective 40×, then the total magnification is 400×.*

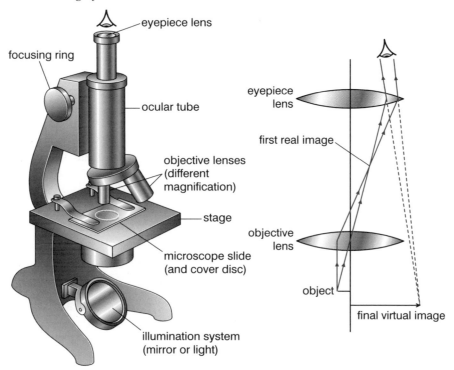

eyepiece lens

focusing ring

ocular tube

eyepiece lens

first real image

objective lenses
(different
magnification)

stage

microscope slide
(and cover disc)

objective lens

object

final virtual image

illumination system
(mirror or light)

COMPOUND MICROSCOPE

electron microscope An electron microscope is a bulky machine which uses a beam of electrons to focus images instead of a beam of light. *A light microscope may magnify 1500 times, but an electron microscope can magnify up to 1 000 000 times.*

electron micrograph An electron micrograph is a photograph of the image produced by an electron microscope.

resolving power is the ability of a microscope to separate two objects which are close together. *It will depend not only on the quality of lenses but also the wavelength used. The shorter the wavelength, the higher the resolving power.*

water and detergents

water is a colourless, tasteless, transparent liquid with highly unusual properties, and is essential to life. *It boils at exactly 100°C when pure, and freezes at 0°C to form ice which is (unusually) less dense than the liquid form. It can be identified in the laboratory by adding it to anhydrous copper sulphate (which turns white to blue) or anhydrous cobalt chloride (which turns blue to pink).*

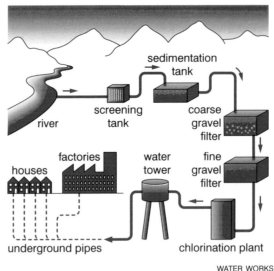

WATER WORKS

chlorination is a method of purifying water by bubbling small amounts of **chlorine** gas through it to kill bacteria,

fluoridation is the addition to drinking water of small amounts of fluoride salts, usually sodium fluoride NaF or calcium fluoride CaF_2. *These fluoride salts combine with the enamel (calcium phosphate) on the **teeth** to form a more protective coating (calcium fluorophosphate). This helps prevent tooth decay.*

hard water is water that is difficult to lather with soap because of dissolved calcium and magnesium ions in the water. *These ions react with soap to form a scum, and also fur up kettles and hot-water pipes.*

soft water is water that is easy to lather with soap as it has few dissolved calcium and magnesium ions. *The 'softest' water of all is **distilled water** and 'deionised water'.*

temporary hardness is hardness which can be removed on heating. *It is due to dissolved calcium (or magnesium) hydrogen carbonate which on heating decomposes to calcium (or magnesium) carbonate. This is*

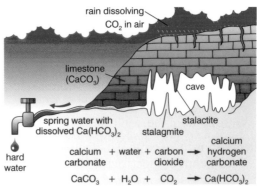

calcium + water + carbon → calcium
carbonate   dioxide   hydrogen carbonate

$$CaCO_3 + H_2O + CO_2 \rightarrow Ca(HCO_3)_2$$

FORMATION OF HARD WATER

Hard water	
Advantages	**Disadvantages**
pleasant taste	needs more soap
good for teeth and bones	leaves messy scum
evidence that it helps prevent heart disease	leaves fur, scale in pipes and kettles

deposited as a scale or fur on the heating element.

calcium hydrogen carbonate	heat $\rightarrow$	calcium carbonate	+	water	+	carbon dioxide
$Ca(HCO_3)_2(aq)$	$\rightarrow$	$CaCO_3(s)$	+	$H_2O(l)$	+	$CO_2(g)$

permanent hardness is hardness which cannot be removed by heating. *It is caused by dissolved calcium or magnesium sulphate ($CaSO_4$ or $MgSO_4$).*

stalactite A stalactite is a downward projection from the ceiling of the roof of a limestone cave. *It is formed from dripping water containing dissolved calcium hydrogen carbonate. The water evaporates leaving a deposit of calcium carbonate. Stalactites may take thousands of years to grow and can eventually join with a **stalagmite** to form a column.*

stalagmite A stalagmite is an upward projection from the floor of a limestone cave.

ion exchange resins remove hardness by exchanging soft sodium ions in the resin for the hard calcium and magnesium ions.

washing soda (or **sodium carbonate**) acts as a water softener and is a cheap method of removing all hardness. *It does this by chemical reaction to precipitate out the calcium and magnesium ions as carbonates.*

detergents (or **soapless detergents** or **synthetic detergents**) are substances added to water to improve its cleaning properties by helping it to dissolve grease. *A detergent molecule is a large molecule with a **covalent** 'tail' and an **ionic** 'head'. The covalent tail is a long hydrocarbon chain which dislikes water (hydrophobic) and attaches itself to grease. The ionic head is attracted to water (hydrophilic). Detergents do not form scum with hard water as their calcium salts are water-soluble. They are manufactured from by-products of the **refining** of petroleum. Detergents should be **biodegradable** and often contain **enzymes** to help break down stains like blood, sweat, etc.*

action of a detergent Detergent molecules attach to grease so that the covalent hydrocarbon tail is buried in the grease or dirt and the ionic head is in the water. *Agitation frees the grease or dirt from the fibres of the material and forms tiny suspended particles surrounded with detergent molecules. Detergent molecules lower the **surface tension** of the water and help the water to thoroughly wet the material. They also help to emulsify fats and oils into solution.*

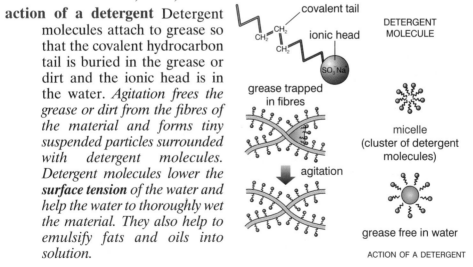

ACTION OF A DETERGENT

257

wave A wave is a regular periodic disturbance in a medium or space. *When an object disturbs the **medium** or space around it the disturbance travels away from the object or source in the form of waves.*

medium A medium is a substance through which a wave travels. *A wave does not permanently disturb the medium through which it travels. A water wave travels across the surface but the water particles do not travel with the wave. They vibrate up and down and remain in their original position.*

progressive wave A progressive wave is one which transports energy (but not matter) away from a source. *Sound and **electromagnetic waves** are progressive waves.*

stationary wave (or **standing wave**) A stationary wave is one which does not transport energy away from a source. *Stationary waves are caused by **interference**.*

transverse wave A transverse wave is a **progressive wave** in which the oscillation or vibration is at right angles to the direction in which the wave is travelling (direction of energy movement). *Only transverse waves can undergo **polarisation**. Examples of transverse waves are **water waves** (oscillation of water molecules) and **electromagnetic waves** (oscillation of electric and magnetic fields).*

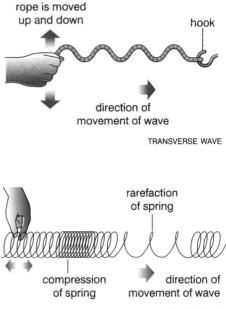

rope is moved up and down

hook

direction of movement of wave

TRANSVERSE WAVE

rarefaction of spring

compression of spring

direction of movement of wave

LONGITUDINAL WAVE

longitudinal wave A longitudinal wave is a **progressive wave** one in which the oscillation or vibration is along the line of the direction in which the wave is travelling (direction of energy movement). *Mechanical waves such as **sound** are longitudinal waves. As the oscillating object moves forward it squashes the particles in the medium together (compressions). When the object moves backwards, the particles in the medium become widely spaced (rarefactions).*

amplitude (a) is the height of a wave from its peak to its mean rest position. *The size of the amplitude indicates the energy carried by the wave and represents, e.g. the loudness of a sound or the brightness of a light.*

wavelength (λ) is the distance between two identical points on the wave, e.g. two adjacent peaks or two adjacent troughs.

frequency (f) is the number of complete waves produced in one second (measured in **hertz**).

period (*T*) is the time of one oscillation (one complete wave). *Period is measured in seconds and is the reciprocal of frequency.*

velocity (*v*) is the distance travelled by a wave in one second. *It is calculated by multiplying the number of complete waves made in one second (**frequency**) by the length of each wave (**wavelength**).*

wave intensity is a measurement of the energy carried by a wave. *It depends on both **amplitude** and **frequency**.*

wave equation The wave equation relates **velocity** (*v*), **frequency** (*f*), and **wavelength** (λ):

$$v = f \times \lambda$$

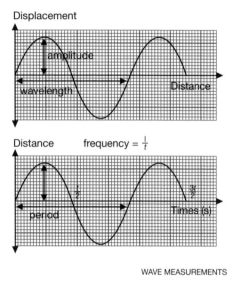

WAVE MEASUREMENTS

hertz (symbol: **Hz**) (or **cycles per second, c.p.s.**) is the SI unit of frequency. *For example, 10 Hz means 10 oscillations per second, producing 10 complete waves per second or 10 complete cycles per second (cps).*

ripple tank A ripple tank is a shallow tank of water with a lamp above which casts shadows on a piece of paper underneath the tank. It is used to study the behaviour of waves. *A vibrating bar or ball produces straight waves or circular waves respectively.*

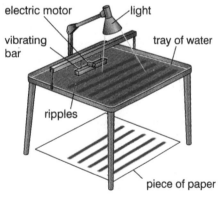

RIPPLE TANK

reflection is the bouncing off of a wave from a barrier. *Only the direction of the wave changes.*

refraction is the change in direction of a wave as it passes from one medium to another. *During refraction there is a change in speed, direction, and wavelength but not in frequency.*

diffraction is the spreading of waves which occurs when a wave goes around an obstacle or through a gap. *There is only a change in direction, not in velocity, frequency, or wavelength.*

interference is the interaction of two or more waves of the same frequency emitted from coherent sources. *If the waves are in phase they may reinforce one another (constructive interference). If they are out of phase they tend to cancel each other out (destructive interference).*

acid rain As a result of acid rain, some lakes in Sweden have become 100 times more acidic than they should be. The water is more acidic than unsweetened orange juice. (see **pollution**)

bacteria Certain species of bacteria are the toughest living organisms known. One species has been found thriving in vents near undersea volcanoes at temperatures of 300°C. Another species of bacterium can withstand up to 10 000 times the human lethal dosage of radiation. (see **bacteria and viruses**)

childbirth The most children born to one woman is sixty-nine. A Russian woman between 1725 and 1765 had sixteen pairs of twins, seven sets of triplets, and four sets of quadruplets. She herself lived for 75 years, outliving many of her children. (see **human reproduction**)

density The heaviest solid is the element osmium, which has a density of $22.5\,\mathrm{g\,cm^{-3}}$ (more than twice as dense as lead). The lightest gas is hydrogen with a density of $0.00009\,\mathrm{g\,cm^{-3}}$. (see **density**)

enzymes It has been estimated that in the human body there are over 7000 different enzymes catalysing various biological changes. So far only about 5000 have been identified. (see **enzymes**)

food The longest recorded survival without food or water is 18 days by a young Austrian man. He was arrested on the 1st April 1979 and put in a police cell but was totally forgotten by the police. He was discovered close to death on the 18th April 1979 having had neither food nor water. (see **food and nutrition**)

gestation period This is the length of pregnancy. The mammal with the longest is the Asiatic elephant: generally around 610 days but up to 760 days. (see **reproduction**)

hair colour The difference in the colour of people's hair is due to the presence of different transition metal compounds. Common brown hair contains compounds of iron, cobalt, and copper. Red hair contains molybdenum compounds and blonde hair contains titanium compounds. (see **transition metals**)

insects There are more different species of insect than any other animal. There are about 1 million species, and about 8000 new species are discovered each year. (see **classification (invertebrates)**)

jacking system The greatest ever lift used 122 hydraulic jacks to raise eight platforms of an offshore oil drilling complex. The lift raised 40 000 tonnes of platform by 6.5 metres (21 feet). (see **machines**)

kilowatt A single 1 kW bar of an electric fire will run on the energy from one kilogram of coal for 2 hours or one kilogram of oil for 4 hours. However it will run on the energy of one kilogram of uranium fuel for 44 000 hours (about 5 years). (see **energy, work and power**)

lightning Lightning strikes the Earth on average 100 times every second. American Roy Sullivan was struck a total of 7 times between 1942 and 1977 and became known as the 'human lightning conductor of Virginia'. (see **electricity (statics)**)

molecules Suppose you decided to share a glass of water ($54\,\mathrm{cm^3}$) with the 6000 million people in the world today. Each man, woman, and child would receive 300 million million molecules of water each. However, nobody would even taste it. (see **moles**)

nerve gas The most powerful nerve gas is three hundred times more poisonous than arsenic. With a chemical name of ethyl-2-diisopropylaminoethylmethylphosphonothiolate, it would send anybody to sleep! (see **organic chemistry**)

ova The animal which lays the largest number of eggs or ova is the Ocean Sunfish. It may lay up to 30 000 000 at one time. However, less than a dozen will become adult fish. (see **reproduction**)

pollution At Sudbury in Canada, the vegetation around the nickel smelters was totally devastated by sulphur dioxide emissions. It made the landscape so desolate that the American Space Agency, NASA, practised their moonlanding procedures there. (see **pollution**)

quark This is a fundamental particle of all atoms. Unlike protons or electrons, quarks have fractions of electronic charge ($+\frac{2}{3}$ or $-\frac{1}{3}$). The proton consists of three quarks , two 'ups' and one 'down': $\frac{2}{3} + \frac{2}{3} - \frac{1}{3} = 1$. (see **atomic structure**)

radioactive blankets Fifty years ago 'radioactive blankets' were sold to help cure arthritis in the legs. The radiation given off penetrated the leg tissue and supposedly cured the disease. Nowadays such blankets would be regarded as highly dangerous. (see **radioactivity**)

smell The animal which has the most acute sense of smell is the male Emperor moth. Using its antennae it can detect a female Emperor moth upwind for up to 11 kilometres. (see **sensitivity**)

transpiration A large mature oak tree would transpire around 150 litres of water through the underneath of its leaves on a warm sunny day. This is equivalent to leaving a powerful water sprinkler running full on for about 10 minutes. (see **plants**)

unmanageable hair When you shampoo your hair the alkali in the soap causes the small scales on each hair to open out. Hair conditioner is a mild acid which neutralises the alkali and causes the scales to close up. Squeezing a lemon over your hair would have the same effect. (see **acids, bases, alkalis**)

velocity The fastest moving molecule is that of hydrogen gas which at 25°C moves around at a speed of approximately 6880 km/h. (see **motion**)

water 2% of the world's water is in the form of ice, and 98% is in the form of liquid water. Only a negligible 0.0004% is in the form of water vapour. (see **water**)

xenotransplants Organs which are transplanted from a member of one species into another. Recently organs from pigs have been successfully transplanted into humans. (see **urinary system**)

yolk The greatest claim for the number of yolks in a chicken's egg is nine. This was reported in 1971 from a poultry farm in New York. (see **reproduction**)

zip The world's longest zip-fastener is used to cover an aquatic cable in Italy. It has 119 007 nylon teeth.

OXFORD

UNIVERSITY PRESS

Great Clarendon Street, Oxford OX2 6DP

Oxford University Press is a department of the University of Oxford.
It furthers the University's objective of excellence in research, scholarship,
and education by publishing worldwide in

Oxford New York

Auckland Cape Town Dar es Salaam Hong Kong Karachi
Kuala Lumpur Madrid Melbourne Mexico City Nairobi
New Delhi Shanghai Taipei Toronto

With offices in

Argentina Austria Brazil Chile Czech Republic France Greece
Guatemala Hungary Italy Japan Poland Portugal Singapore
South Korea Switzerland Thailand Turkey Ukraine Vietnam

Oxford is a registered trade mark of Oxford University Press
in the UK and in certain other countries

Text © Chris Prescott 1999

The moral rights of the author have been asserted

Database right Oxford University Press (maker)

First published 1996
Second edition 2006
Third edition 2008
This edition 2013

British Library Cataloguing in Publication Data

Data available

ISBN 9780 19 273358 0

5 7 9 10 8 6 4

Printed in Singapore by KHL Printing Co.Pte Ltd

Paper used in the production of this book is a natural, recyclable product
made from wood grown in sustainable forests. The manufacturing process conforms to
the environmental regulations of the country of origin.

www.oxforddictionaries.com/schools